Lecture Notes in Computer Science 10612

Commenced Publication in 1973
Founding and Former Series Editors:
Gerhard Goos, Juris Hartmanis, and Jan van Leeuwen

Editorial Board

More information about this series at http://www.springer.com/series/7409

Shaoxu Song · Matthias Renz
Yang-Sae Moon (Eds.)

Web and Big Data

APWeb-WAIM 2017 International Workshops: MWDA,
HotSpatial, GDMA, DDC, SDMA, MASS
Beijing, China, July 7–9, 2017
Revised Selected Papers

Springer

Editors
Shaoxu Song
Tsinghua University
Beijing
China

Matthias Renz
George Mason University
Fairfax, VA
USA

Yang-Sae Moon ⓘ
Kangwon National University
Chuncheon
Korea (Republic of)

ISSN 0302-9743   ISSN 1611-3349   (electronic)
Lecture Notes in Computer Science
ISBN 978-3-319-69780-2   ISBN 978-3-319-69781-9   (eBook)
https://doi.org/10.1007/978-3-319-69781-9

Library of Congress Control Number: 2017957820

LNCS Sublibrary: SL3 – Information Systems and Applications, incl. Internet/Web, and HCI

Printed on acid-free paper

This Springer imprint is published by Springer Nature
The registered company is Springer International Publishing AG
The registered company address is: Gewerbestrasse 11, 6330 Cham, Switzerland

Preface

The Asia Pacific Web (APWeb) and Web-Age Information Management (WAIM) Joint Conference on Web and Big Data is a leading international conference for researchers, practitioners, developers, and users to share and exchange their cutting-edge ideas, results, experiences, techniques, and tools in connection with all aspects of Web data management. As the first joint event, APWeb-WAIM 2017 was held in Beijing, China, during July 7–9, 2017, and it attracted participants from all over the world.

Along with the main conference, APWeb-WAIM workshops intend to provide an international forum for researchers to discuss and share research results. This APWeb-WAIM 2017 workshop volume contains the papers accepted for the following six workshops that were held in conjunction with APWeb-WAIM 2017. These six workshops were selected after a public call for proposals process, each of which focuses on a specific area that contributes to the main themes of the APWeb-WAIM conference. The six workshops were as follows:

- The International Workshop on Mobile Web Data Analytics (MWDA 2017)
- The International Workshop on Hot Topics in Big Spatial Data and Urban Computing (HotSpatial 2017)
- The International Workshop on Graph Data Management and Analysis (GDMA 2017)
- The Second International Workshop on Data-Driven Crowdsourcing (DDC 2017)
- The Second International Workshop on Spatio-temporal Data Management and Analytics (SDMA 2017)
- The International Workshop on Mobility Analytics from Spatial and Social Data (MASS 2017)

We would like express our thanks to all the workshop organizers and Program Committee members for their great effort in making the APWeb-WAIM 2017 workshops a success. In total, 25 papers were accepted for the workshops. In particular, we are grateful to the main conference organizers for their generous support and help.

August 2017

Shaoxu Song
Matthias Renz
Yang-Sae Moon

Organization

MWDA 2017

Workshop Chairs

Li Li	Southwest University, China
Li Liu	Chongqing University, China
Xiangliang Zhang	King Abdullah University of Science and Technology, Saudi Arabia

Program Committee

Shiping Chen	CSIRO, Australia
Jiong Jin	Swinburne University of Technology, Australia
Ming Liu	Southwest University, China
Guoxin Su	National University of Singapore, Singapore
Min Gao	Chongqing University, China
Rong Xie	Wuhan University, China
Huawen Liu	Zhejiang Normal University, China
Lifei Chen	Fujian Normal University, China
Basma Alharbi	King Abdullah University of Science and Technology, Saudi Arabia
Zehui Qu	Southwest University, China
Yonggang Lu	Lanzhou University, China
Zhen Dong	National University of Singapore, Singapore
Ye Liu	National University of Singapore, Singapore
Jun Zeng	Chongqing University, China
Aiguo Wang	Hefei University of Technology, China
Xuejun Li	Anhui University, China
Chenren Xu	Peking University, China
Danni Wang	Chongqing University, China

HotSpatial 2017

Workshop Chairs

Huy T. Vo	City University of New York, USA
Weixiong Rao	Tongji University, China

Program Committee

Huy T. Vo	City University of New York, USA
Kai Zhao	New York University, USA
Yu Xiao	Aalto University, Finland
Liang Wang	University of Cambridge, UK
Jia Zeng	Huawei Hong Kong, SAR China
Yong Li	Tsinghua University, China
Weixiong Rao	Tongji University, China
Mingxuan Yuan	Huawei, Hong Kong, SAR China
Yi Cai	South China University of Technology, China
Qinpei Zhao	Tongji University, China
Aaron Yi Ding	TU Munich, Germany

GDMA 2017

Workshop Chairs

Lei Zou	Peking University, China
Xiaowang Zhang	Tianjin University, China
Lijun Chang	University of New South Wales, Australia
Zhiwei Zhang	Hong Kong Baptist University, SAR China

Program Committee

Robert Brijder	Hasselt University, Belgium
G.H.L. Fletcher	Technische Universiteit Eindhoven, The Netherlands
Dirk Van Gucht	Indiana University, USA
Jelle Hellings	Hasselt University, Belgium
Liang Hong	Wuhan University, China
Egor V. Kostylev	University of Oxford, UK
Xueming Lin	The University of New South Wales, Australia
Shuai Ma	Beihang University, China
Peng Peng	Hunan University, China
Sherif Sakr	University of New South Wales, Australia
Dennis Shasha	New York University, USA
Hongzhi Wang	Harbin University of Industry, China
Junhu Wang	Griffith University, Australia
Kewen Wang	Griffith University, Australia
Guohui Xiao	Free University of Bozen-Bolzano, Italy
Xifeng Yan	University of California at Santa Barbara, USA
Jeffrey Xu Yu	Chinese University of Hong Kong, SAR China
Ye Yuan	Northeast University, China
Weiguo Zheng	The Chinese University of Hong Kong, SAR China
Lu Qin	University of Technology Sydney, Australia

Yuanyuan Zhu	Wuhan University, China
Xin Huang	Hong Kong Baptist University, SAR China
Ronghua Li	Shen Zhen University, China
Zechao Shang	The University of Chicago, USA

DDC 2017

Workshop Chairs

Longfei Shangguan	Princeton University, USA
Zhiyang Su	Microsoft, China
Yongxin Tong	Beihang University, China
Wenjun Wu	Beihang University, China

Program Committee

Caleb Chen Cao	Financial Data Technology Ltd., Hong Kong, SAR China
Yuqiang Chen	4Paradigm Inc., China
Yurong Cheng	Northeast University, China
Bolin Ding	Microsoft Research, USA
Ju Fan	Renmin University of China, China
Xiaonan Guo	Stevens Institute of Technology, USA
Guoliang Li	Tsinghua University, China
Hailong Sun	Beihang University, China
Xike Xie	University of Science and Technology of China, China
Hongzhi Yin	University of Queensland, Australia
Dongxiang Zhang	University of Electronic Science and Technology of China, China
Chen Jason Zhang	Hong Kong University of Science and Technology, SAR China
Xiaolong Zheng	Tsinghua University, China
Zimu Zhou	ETH Zurich, Switzerland

SDMA 2017

Workshop Chairs

Xiaoyong Du	Renmin University of China
Xiaofang Zhou	University of Queensland, Australia
Lei Chen	Hong Kong University of Science and Technology, SAR China
Kuien Liu	Alibaba Inc.
Wei Lu	Renmin University of China

Program Committee

Zhiming Ding	Beijing University of Technology, China
Yunjun Gao	Zhejiang University, China
Hao Huang	Wuhan University, China
Xiaohui Hu	Institute of Software, Chinese Academy of Sciences, China
Peiquan Jin	University of Science and Technology of China
Zhixu Li	Soochow University, Taiwan
Chengfei Liu	Swinburne University of Technology, Australia
Jie Liu	Institute of Software, Chinese Academy of Sciences, China
Feng Lu	IGSNRR, Chinese Academy of Sciences, China
Hua Lu	Aalborg University, Denmark
Yuwei Peng	Wuhan University, China
Weiwei Sun	Fudan University, China
Guangzhong Sun	University of Science and Technology of China
Gang Pan	Zhejiang University, China
Chuitian Rong	Tianjin Polytechnic University
Shuo Shang	China University of Petroleum-Beijing
Shaoxu Song	Tsinghua University, China
Yanyan Shen	Shanghai Jiaotong University, China
Huanliang Sun	Shenyang Jianzhu University
Sai Wu	Zhejiang University, China
Xing Xie	Microsoft Research Asia
Jiajie Xu	Soochow University, Taiwan
Jianqiu Xu	Nanjing University of Aeronautics and Astronautics, China
Zheng Xu	The Third Research Institute of the Ministry of Public Security, China
Rui Yang	Tsinghua University, China
Yang Yue	Shenzhen University, China
Chang Yao	National University of Singapore, Singapore
Pengpeng Zhao	Soochow University, Taiwan
Meihui Zhang	Singapore University of Technology and Design, Singapore
Wen Zhang	Beijing University of Chemical Technology, China
Dongxiang Zhang	University of Electronic Science and Technology of China

MASS 2017

Workshop Chairs

Qiang Qu	Shenzhen Institutes of Advanced Technology, Chinese Academy of Sciences, China
Shuhui Wang	Institute of Computing Technology, Chinese Academy of Sciences, China
Siyuan Liu	Pennsylvania State University, USA
JooYoung Lee	Innopolis University, Russia

Program Committee

Hongyan Li	Peking University, China
JooYoung Lee	Innopolis University, Russia
Laura Radaelli	Tel Aviv University, Israel
Leon Derczynski	The University of Sheffield, UK
Yongluan Zhou	University of Southern Denmark
Jiang Qingshan	SIAT
Chulaka Gunasekara	IBM Research

Contents

GDMA 2016

DDC 2017

SDMA 2017

MASS 2017

MWDA 2017

Smooth Representation Clustering Based on Kernelized Random Walks

Liping Chen$^{(\boxtimes)}$, Gongde Guo, and Lifei Chen

School of Mathematics and Informatics, Fujian Normal University,
Fuzhou 350007, People's Republic of China
{lpchen, ggd, clfei}@fjnu.edu.cn

Abstract. With the widespread use of smart phones and tablet computers, it is necessary to develop algorithms to assist high throughout analysis of mobile videos. A novel method for automated segmentation on the mobile video scenery is proposed in this paper. It uses the kernelized random walks on the globe KNN graph and the Smooth Representation Clustering to improve the segmentation effectiveness. The high order transition probability matrix of the kernelized random walks is utilized for erasing the unreliable edge of the graph. Simultaneously kernel approach is used to assign different weights for neighbors to evaluate their contribution to the clustering. The method is evaluated on two public datasets and a real-world mobile video taken by a smart phone. The experimental results show that the proposed algorithm achieves better performance compared with the other representative algorithms.

Keywords: Subspace clustering · Smooth Representation Clustering · The random walks · KNN graph

1 Introduction

With the rapid development of Mobile Internet, mobile devices such as smart phones and small pads have greatly changed people's way of life in recent years. From the ericsson mobility report 2014, 90% of population all over the world will hold a smart phone by 2020, whenever people like and wherever people prefer they can easily take photos and even record videos and share them in Youku, Tencent, WeChat and another real-time sharing system via mobile devices. The mobile videos are very rich in daily life resources and become the major style of entertainment during people's spare time [1].

Such kind of data in the mobile platform often lie in a mixture of low dimensional subspaces, each subspace belonging to one class or cluster. Hence, the work to classify scene on the mobile video automatically grows in importance [2]. The aim of the proposed algorithm is to solve the work based on Smooth Representation Clustering and Kernelized Random Walks (RSMR).

S. Song et al. (Eds.): APWeb-WAIM 2017 Workshops, LNCS 10612, pp. 3–10, 2017.
https://doi.org/10.1007/978-3-319-69781-9_1

The contribution of our paper lies in:

(1) The kernelized random walk theory is applied aimed to kick off these unbeliev-able neighbors in KNN graph.

(2) In normal KNN method all neighbors are considered to give equal donations to the clustering of the testing point. We use the kernel function as the similarity measure to assign different weights for neighbors to evaluate their contribution to the clustering.

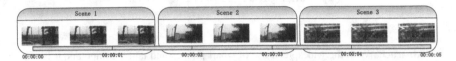

Fig. 1. The scene clustering on the mobile video

2 Related Works

2.1 Subspace Clustering

The subspace clustering of mobile videos is an emerging requirement in real life. Many useful tasks such as scenery clustering and motion segmentation are needed. These tasks are highly integrated with professional technology and actual demand. As a result, people can comfortably enjoy the convenience of technology without perceiving the complexity of technology. This subspace clustering process is divided into two steps: The first step is to establish a global sparse optimization model in which the high dimensional data are represented with the linear combination of the same subspace in the low dimension to compute affinities; the second step is to apply the sparse coef-ficient matrix to a new spectral clustering algorithm framework. So the key of the spectral clustering is to construct the affinity matrix. Elhamifar [3–5] proposed a sparse subspace clustering model based on the data's self-expression. A similar framework [6, 7] is used to solve the structure of the nuclear norm graph. The algorithm proposed in [7] calculates the joint expression matrix of the nuclear norm to estimate the reconstruction factor from the global structure of the coefficient matrix. Lu et al. [8] proposed an adaptive subspace processing in the optimization of the subspace. Chen et al. [9] and Lu et al. [10] advance the grouping effect based on LSR and CASS framework respectively. In that representation coefficients are similar to each other, which means samples are also close to each other. The grouping effect of representation is extended in depth by the Smooth Representation Clustering (SMR) [11] with elegant formulations as:

$$\min_Z f(Z) = \alpha \|X - XZ\|_F^2 + tr(ZLZ^T) \tag{1}$$

$$tr(ZLZ^T) = \frac{1}{2}\sum_{i=1}^N\sum_{j=1}^N w_{ij}\|z_i - z_j\|_2^2 \tag{2}$$

In the formula the matrix $W = (w_{ij})_{N \times N}$ is the weight matrix which measures the spatial closeness of the samples. But the neighbors in the spatial domain don't belong to the same subspace. As the result the normal 0–1 weighted k-nn graph has some untrustworthy edge. It has been justified in paper [12] these error edges can be effectively reduced by the high order transition probability matrix.

2.2 The Random Walks on the Graph

In 1945, Karl Pearson [13] puts forward the random walks theory to describe the diffusion process. The random walks process on the graph [14] refers to the node move to the other nodes according to the given probability. When the distance between the adjacent nodes is small, at the same time, the similarity between nodes is higher. It means that random walk is more possible to walk along the edge. Consequently the edge weight should be larger as the transition is more likely to happen. General speaking, the edge weight on the graph not only represents the structure of the graph, but also affects the random walks in the graph.

3 The Proposed Algorithm

3.1 The Kernelized Random Walks on the Graph

It is well known that the video sequence captured by the mobile device is a kind of typical time series format data. Though the transition process between the frames is not completely random in fact, but samples wander from origin state i to the next state is random. At this condition the time constraint plays an important role in visual information understanding problems. One of the convenient and effective methods to describe the time correlation characteristics between adjacent images is the Markov random field model. Since the random walks are a special case of Markov random chains, we are ready to apply the random walks model [15–19] maturely. The Outstanding achievements of the nonlinear techniques have been achieved in image processing applications [20–23] By selecting the appropriate kernel map, the sparse representation of the signal in the feature space can be obtained with the random walks and regulation method as it is shown in Fig. 2.

The Kernelization technology is one of the important nonlinear techniques. In this paper we improve the Random Walk model from the perspective of kernelization. Given a sample data set $X = [x_1, x_2, \cdots x_N] \in R^{D \times N}$, assuming we make the nonlinear transform on X, the transform is noted as $\phi(X) = [\phi(x_1), \phi(x_2), \cdots \phi(x_N)]$. The application of kernel technique in random walker model doesn't not need to give nonlinear transform explicitly, so we directly define kernel function $k(x_i, x_j)$ to show the similarity within the samples as $k(X, x_j) = [k(x_1, x_j), k(x_2, x_j), \cdots, k(x_N, x_j)]^T$. So that the kernel matrix K defined on the domain (X, X) is represented as $K = k(X, X) = [k(x_i, x_j)]_{N \times N}$.

If we use the graph to describe the kernelized random walker model, an undirected graph can be noted as <G, E, W> where G is the set of samples, E is the set of the edges and W is the weights of the edges. It is assumed the whole samples set consists of m possible classes and each sample belongs to one of these classes. After we calculate the kernel function between all pairs of samples in order to select the K nearest neighbors, each edge is consisted of the pair of samples which are K nearest neighbors in the graph. The weight matrix of the graph is defined as:

$$W(x_i, x_j) = \begin{cases} 1 & if \quad i \equiv j \\ k(x_i, x_j) = \exp(-d(x_i, x_j)/\omega) & if \quad i \quad and \quad j \quad are \quad knn \quad neighbors \\ 0 & others \end{cases} \quad (3)$$

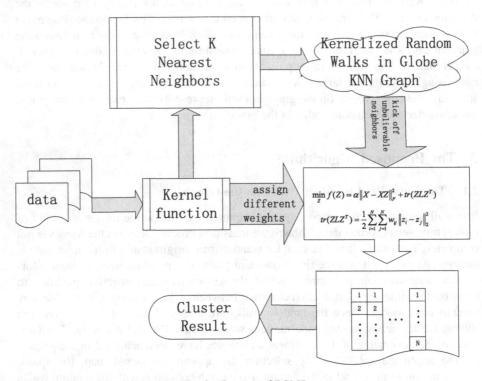

Fig. 2. Process of RSMR

Here $d(X_i, X_j)$ stands for the Euclidean distance, ω is the control factor which value can be tuned according to the distance. Furthermore the Gaussian kernel function is used to convert Euclidean distance to similarity representation conveniently. As the result contributions of the k nearest training data points are weighted according to the similarity between X_i and X_j. In another word the greater weight means the closer kNN data points. When we regard each vertex in the graph as a state, this measure also favors transitions between different states that belong to the same class according to the given distance. When the difference between vertex i and vertex j is small, the larger probability random walks go along the edge from i to j; When the difference between adjacent vertex is bigger, the random walks choose the edge with less probability. So the transition probability of a random walk beginning from sample i arriving at any one sample can be constructed as below:

$$P_w(x_i, x_j) = \frac{W(x_i, x_j)}{\sum\limits_{j=1}^{N} W(x_i, x_j)} \tag{4}$$

The value of $P_w(x_i, x_j)$ lies in the interval [0, 1]. Since the higher order transition probability matrix of the random walker calculated by the Chapman Kolmogorov Equation indicates the more static relationship among the data, we apply the t order transition probability $P_w^t(X_i, X_j)$ to remove these unreliable edges.

$$W'(x_i, x_j) = \begin{cases} W(x_i, x_j) & \text{if} \quad \sum\limits_{j \in (i-\delta, i+\delta)} P_w^t(x_i, x_j)/2\delta > \alpha \ or \quad i \equiv j \\ 0 & others \end{cases} \tag{5}$$

Here the constraint $(i - \delta, i + \delta)$ indicates the neighborhood relationship in the domain. In the image sequence the constraint means the neighborhood in the time domain. α is the threshold, it can select the first K * N/10 largest transition probability value in the row i. As the images of KNN neighbors shown in the Fig. 3, a lot of unreliable KNN neighbors (signed as red dots) are removed on the USPS database.

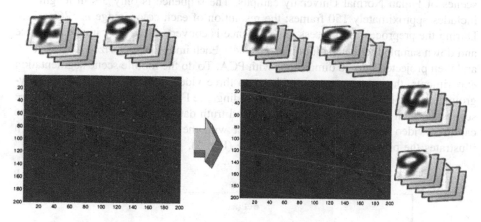

Fig. 3. The partial KNN affinity matrix (Color figure online)

3.2 The Solution and Algorithm

When we take the new weight matrix into the aim formula, the objective function turns to be:

$$\min_Z f(Z) = \alpha\|X - XZ\|_F^2 + tr(Z\tilde{L}Z^T) \tag{6}$$

The matrix Z stands for the coefficient matrix, and the Laplacian matrix is calculated with $\tilde{L} = D - W' + \varepsilon I$. In the computation, εI is added to ensure L to be positive definite and by default $\varepsilon = 0.01$. The answer to the objective function lies in the Sylvrster equation [11] as $\alpha X^T XZ + XZL' = \alpha X^T X$. Afterwards we get m clusters by spectral clustering with the affinity matrix $A = (|Z| + |Z|^T)/2$ or $A = \left| \dfrac{z_i^T z_j}{\|x_i\|_2 \|x_j\|_2} \right|^\gamma$.

4   Experiments

We report numerical results with the Clustering errors (CE) measure [24–26] to evaluate the output results and the ground truth under the optimal variation. In our experiments, we use two datasets and a short video taken by a smart phone to evaluate the performance of our algorithm.

The USPS [26] database is a handwritten digit dataset collected by the American Postal Service. It includes 9298 pictures whose size is 16×16 pixels. We choose the first 100 images of each digit for the experiments as the paper [11] does. The performance on the USPS database is further enhanced again. We attain error rate of 7.3 just better than the other candidates under the same experiment setting.

We also test SSC, LRR, SMR and RSMR approach on a sample video captured by iphone 4 s with qualitative results shown in Fig. 4. The video sequences contain three scenes of Fujian Normal University campus. The sequence is only 5 s in length and includes approximately 150 frames; the resolution of each color image is 576×320. During the preprocessing process, the sequence is converted to a gray image sequence and down sampled to the resolution of 72×40. Each image is vectorized to $X_i \in R^{2880}$ and then projected into 36 dimensions with PCA. To do the mobile scene segmentation experiments, the sequences are divided into three video clips about the school playground, school hospital and the teaching building, see Fig. 1 for more details. The three scenes are then labeled by hand as the ground truth data. RSMC performs best on the captured video sequences in the experiment with the lowest error rate of 6, which illustrates the proposed approach is effective (Fig. 5).

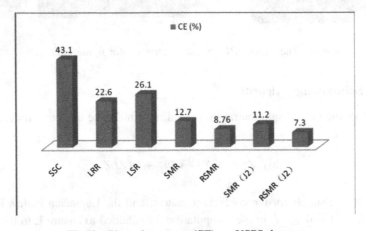

Fig. 4. Clustering errors (CE) on USPS datasets

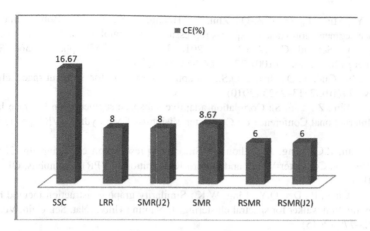

Fig. 5. Clustering errors (CE) On mobile dataset

5 Conclusions

The proposed algorithm is based on random walks on the globe KNN graph and the Smooth Representation Clustering. The higher order transition probability matrix of the kernelized random walks is utilized for erasing the unreliable edge of the graph. We use the kernel function to give different weights for neighbors by evaluating their contribution to the clustering. The approach has been shown to outperform some respective algorithms in the experiments.

Acknowledgments. This work was supported by Chinese National Natural Science Foundation under Grant Nos. 61672157, 41601477, it is also supported by the Leading project in Science and Technology Department of Fujian Province under Grant No. 2015Y0054.

References

1. Wang, J., Xu, M., Lu, H.: ActiveAd: a novel framework of linking ad videos to online products. Neurocomputing **185**, 82–92 (2016)
2. Vidal, R.: A tutorial on subspace clustering. IEEE Signal. Proc. Mag. **28**(2), 52–68 (2011)
3. Elhamifar, E., Vidal, R.: Sparse subspace clustering: algorithm, theory, and applications. IEEE Trans. Pattern Anal. Mach. Intell. **35**(11), 2765–2781 (2013)
4. Elhamifar, E., Vidal, R.: Sparse subspace clustering. In: IEEE Conference of Computer Vision and Pattern Recognition (CVPR 2009), Florida, pp. 2790–2797 (2009)
5. Elhamifar, E., Sapiro, G., Sastry, S.: Dissimilarity-based sparse subset selection. IEEE Trans. Pattern Anal. Mach. Intell. **PP**(99), 1 (2015). doi:10.1109/tpami.2015.2511748(2015)
6. Jun, X., Kui, X.: Reweighted sparse subspace clustering. Comput. Vis. Image Underst. **138**, 25–37 (2015)
7. Liu, G., Lin, Z., Yan, S., Sun, J., Yu, Y., Ma, Y.: Robust recovery of subspace structures by low-rank representation. IEEE Trans. Pattern Anal. Mach. Intell. **35**(1), 171–184 (2012). doi:10.1109/tpami.2012.88

8. Lu, C.-Y., Min, H., Zhao, Z.-Q., Zhu, L., Huang, D.-S., Yan, S.: Robust and efficient subspace segmentation via least squares regression. In: Fitzgibbon, A., Lazebnik, S., Perona, P., Sato, Y., Schmid, C. (eds.) ECCV 2012. LNCS, vol. 7578, pp. 347–360. Springer, Heidelberg (2012). doi:10.1007/978-3-642-33786-4_26

9. Chen, L.F., Guo, G.D., Jiang, Q.S.: Adaptive algorithm for soft subspace clustering. J. Softw. **21**(10), 2513–2523 (2010)

10. Lu, C.Y., Lin., Z., Yan, S.: Correlation adaptive subspace segmentation by trace lasso. In: IEEE International Conference on Computer Vision (ICCV), Sydney, VIC, pp. 1345–1352 (2013)

11. Hu, H., Lin, Z.C., Feng, J.J., Zhou, J.: Smooth representation clustering. In: 2014 IEEE Conference on Computer Vision and Pattern Recognition (CVPR), Columbus, OH, pp. 23–28 (2014)

12. Cao, J.Z., Chen, P., Dai, Q.Y., Ling, W.K.: Similarity graph construction method based on Markov random walker for spectral clustering. J. Nanjing Univ. (Nat. Sci. Chin. Version) **51**(4), 772–779 (2015)

13. Kakutani, S.: Markov processes and the Dirichlet problem. Proc. Jpn. Acad. **21**, 227–233 (1945)

14. İnkaya, T.: A parameter-free similarity graph for spectral clustering. Expert Syst. Appl. **42**(24), 9489–9498 (2015)

15. Grady, L.: Random walks for image segmentation. IEEE Trans. Pattern Anal. Mach. Intell. **28**(11), 1768–1783 (2006)

16. Grady, L.: Multilabel random walker image segmentation using prior models. In: IEEE Conference of Computer Vision and Pattern Recognition (CVPR 2005), San Diego, vol. 1, pp. 763–770 (2005)

17. Grady, L., Sinop, A.K.: Fast approximate random walker segmentation using eigenvector precomputation. In: IEEE Conference of Computer Vision and Pattern Recognition (CVPR 2008), Anchorage, Alaska, pp. 24–26 (2008)

18. Sonu, K.J., Purnendu, B., Subhadeep, B.: Random walks based image segmentation using color space graphs. Proc. Technol. **10**(2013), 271–278 (2013)

19. Yang, X.L., Su, Y., Duan, R.B., Fan, H.J., Yeo, S.Y., Lim, C., Zhang, L., Tan, R.S.: Cardiac image segmentation by random walks with dynamic shape constraint. IET Comput. Vis. **10**(1), 79–86 (2016)

20. Mitr, S.K., Sicuranza, G.L.: Nonlinear image processing, pp. 274–278. Academic Press, Cambridge (2001)

21. Schölkopf, B., Smola, A., Müller, K.R.: Nonlinear component analysis as a kernel eigenvalue problem. Neural Comput. **10**(5), 1299–1319 (1998)

22. Ramponi, G., Strobel, N.K., Mitra, S.K., Yu, T.H.: Nonlinear unsharp masking methods for image contrast enhancement. J. Electron. Imaging **5**(3), 353–366 (1996)

23. Li, H., Adali, T.: Complex-valued adaptive signal processing using nonlinear functions. EURASIP J. Adv. Sign. Process. **2008**, 122 (2008)

24. Tron., R., Vidal, R.: A benchmark for the comparison of 3-d motion segmentation algorithm. In: IEEE Conference of Computer Vision and Pattern Recognition (CVPR 2007), Minneapolis, Minnesota, pp. 1–8 (2007)

25. Kanatani, K., Sugaya., Y.: Multi-state optimization for multi-body motion segmentation. In: Proceedings of Australia-Japan Advanced Workshop on Computer Vision, Adelaide Australia, pp. 25–31 (2003)

26. Hull, J.J.: A database for handwritten text recognition research. IEEE Trans. Pattern Anal. Mach. Intell. **16**(5), 550–554 (1994)

A Joint Model for Water Scarcity Evaluation

Jingyuan Wang and Li Li[✉]

School of Computer and Information Science,
Southwest University, Chongqing, China
wjykim@email.swu.edu.cn, lily@swu.edu.cn

Abstract. To make a real difference for our thirsty planet, we establish
the water demand-supply model and the Advanced Water Poverty Index
(AWPI). First, we develop a dynamic demand-supply model to measure
the ability of a region to satisfy its water consumption. On the demand
side, we fit agricultural and industrial water needs by the Grey Verhulst
prediction model, then we consider domestic needs through the Logistic
Growth model of total population and the Regression model of residential
needs per capita. On the supply side, we estimate the impacts of multiple
factors such as utilized internal river and rainfall, desalinated seawater
and purified sewage. In the experiments, we use the sensor data from
the World Bank. Also, the stability of our model has been proved by the
evaluation. Second, we analyze the types of water scarcity by improving
the Water Poverty Index to the Advanced Water Poverty Index, and we
creatively add population as the sixth key component. The prediction
can be used as an important indicator for the government to take some
specific intervention measures to help alleviate the severe water shortage
and achieve sustainable development of water resources.

Keywords: Water scarcity · Grey Verhulst prediction model · Logistic
Growth model · AWPI

1 Introduction

Lack of sufficient available water resources to meet living and production needs,
1.6 billion people are suffering from water scarcity [8,9]. Therefore, all the countries are taking strategies to guard our thirsty planet.

Extracting more and more clean water for agricultural, industrial and residential purposes heavily threats the health of aquatic ecosystems and the life
those ecosystems support. And the explosive population growth is exacerbat-
ing this matter. How to thoroughly analyze the degree and the causes of water
scarcity has become a worldwide issue attracting more and more concern [5,7].

We make some necessary assumptions (for simplifying a realistic model) as
follows.

- We only take the water demand of industry, agriculture and domestic into
  consideration.
- The consumption of water is determined and can be measured by statistic
  method.

© Springer International Publishing AG 2017
S. Song et al. (Eds.): APWeb-WAIM 2017 Workshops, LNCS 10612, pp. 11–20, 2017.
https://doi.org/10.1007/978-3-319-69781-9_2

– The country or region chosen is relatively stable.

Our contributions are as follows.

– **Modeling the water demand and supply to evaluate the scarcity without intervention.** For one thing, we develop the Grey Verhulst prediction model, the Logistic Growth model and the Regression model to analyze agricultural, industrial and residential water demand. For another, we estimate the water supply through utilized internal river and rainfall, desalinated seawater and purified sewage.
– **Analyzing the causes of water scarcity.** To distinguish the types of water scarcity, we establish the Advanced Water Poverty Index (AWPI), in which we consider resource, access, capacity, use, environment, and population as six key components.

2 Notations

In this section, we show the variables and parameters to be used in our work. They are fundamental and vital to the models in the following sections.

We list the variables and parameters used for constructing the models as follows. Notations and their definitions are shown in Table 1.

Table 1. Notations used in the paper

Symbol	Definition	Unit
$r(x)$	The population growth rate	Unitless
x_m	The maximum population which natural resources and environmental conditions can accommodate	Unitless
D_t, D_{at}, D_{it}	Dynamic total, agricultural and industrial water demand over time	Billion cubic meters
U_{dt}, P_{dt}	Dynamic residential water demand per capita and total population over time	Billion cubic meters, unitless
S_t	Dynamic water supply over time	Billion cubic meters
N	Natural water source capacity	Billion cubic meters
A_t, D_{kt}	Dynamic utilization rate of natural water source and sewage of k industry over time	Unitless
SE_t, p_t	Dynamic desalinated seawater and water pollution over time	Billion cubic meters
R_k	Purification rate of k industry	Unitless
ρ	Relative supply-demand ratio	Unitless
RI, AI, CI, UI, EI, PI	Resource, access, capacity, use, environment and population index	Unitless

3   The Demand-Supply Model

3.1   The Demand Model

To analyze the condition of water demand, we must consider the agricultural consumption, industrial consumption and domestic consumption, respectively. Because the monotonically increasing sequence corresponds to the GM (1, 1) model while S-shaped sequence corresponds to the Grey Verhulst prediction model [1,13], we randomly choose five countries to judge the shape of their agricultural and industrial usage to match a proper model. As for the domestic usage, we develop the Logistic Growth model [12] to predict the population growth, and we establish the Regression model to predict the domestic water consumption per capita.

3.1.1   The Grey Verhulst Prediction Model

First, we use the annual freshwater withdrawals, industry/agriculture from World Bank database[1] (Romania, Cyprus, Iraq, India and China) (Fig. 1), which apparently justifies a non-monotonic increase; thus we establish the Grey Verhulst prediction model instead of the GM (1, 1) Model.

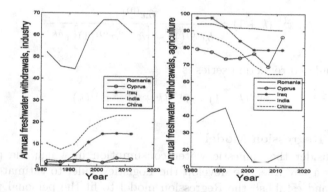

Fig. 1. The annual freshwater withdrawals, industry/agriculture

We select a sequence

$$x^{(0)} = x^{(0)}(1), x^{(0)}(2), ..., x^{(0)}(n) \tag{1}$$

and we can get another sequence within the following relationship

$$\begin{aligned} x^{(1)} &= x^{(1)}(1), x^{(1)}(2), \ ., x^{(1)}(n) \\ &= x^{(0)}(1), x^{(0)}(1) + x^{(0)}(2), ..., x^{(0)}(1) + ... + x^{(0)}(n) \end{aligned} \tag{2}$$

---

[1] http://data.worldbank.org.

where $x^{(1)}(k) = \sum_{i=1}^{k} x^{(0)}(i)$, $k = 1, 2, ..., n$

Also,

$$z^{(1)} = z^{(1)}(2), z^{(1)}(3), ..., z^{(1)}(n) \tag{3}$$

where the correction among these variables is

$$z^{(1)}(k) = 0.5x^{(1)}(k) + 0.5x^{(1)}(k-1), \quad k = 2, 3, ..., n \tag{4}$$

We then obtain $x^{(0)} + az^{(1)} = b(z^{(1)})^2$ as well as the differential equation $\frac{dx^{(1)}}{dt} + ax^{(1)} = b(x^{(1)})^2$.

To estimate the parameters, we make

$$u = [a, b]^T, B = \begin{bmatrix} -z^{(1)}(2) \ (z^{(1)}(2))^2 \\ -z^{(1)}(3) \ (z^{(1)}(3))^2 \\ \vdots \qquad \vdots \\ -z^{(1)}(n) \ (z^{(1)}(n))^2 \end{bmatrix}, Y = \begin{bmatrix} x^{(0)}(2) \\ x^{(0)}(3) \\ \vdots \\ x^{(0)}(n) \end{bmatrix} \tag{5}$$

then we get that $\hat{u} = [\hat{a}, \hat{b}]^T = (B^T B)^{-1} B^T Y$ though least squares method.

Therefore the solution to the differential equation is

$$\hat{x}^{(1)}(k+1) = \frac{\hat{a}x^{(0)}(1)}{\hat{b}x^{(0)}(1) + \left[\hat{a} - \hat{b}x^{(0)}(1)\right] e^{\hat{a}k}} \tag{6}$$

so the accumulated reduction series is

$$\hat{x}^{(0)}(k+1) = \hat{x}^{(1)}(k+1) - \hat{x}^{(1)}(k) \tag{7}$$

3.1.2   The Regression Model

Then we consider the domestic water consumption in which the population growth plays a key role. We develop the Logistic model to estimate the population size and establish the Regression model to fit the personal water consumption.

The population growth rate varies as $r(x) = r(1 - \frac{x}{x_m})$ where r is a constant, x is the population, and x_m is the maximum population which natural resources and environmental conditions can accommodate.

Then, we get the ordinary differential equations

$$\begin{cases} \frac{dx}{dt} = r(1 - \frac{x}{x_m})x \\ x(t_0) = x_0 \end{cases} \tag{8}$$

and its solution

$$x(t) = \frac{x_m}{1 + (\frac{x_m}{x_0} - 1)e^{-r(t-t_0)}} \tag{9}$$

Third, we use the Regression model to fit the domestic water consumption per capita. As a benchmark model, we simply assume the time series data meets a linear relationship. It's rational because the difference between this fundamental model and the improved one just exists in the complexity of a function.

The dependent variable y represents domestic water consumption per capita (cubic meter/person), and the independent variable x represents time (year).

We assume that

$$y = \beta_0 + \beta_1 x \tag{10}$$

Then we calculate the estimates of the parameters by minimizing the differences between data points and the line:

$$\begin{cases} \hat{\beta}_0 = \bar{y} - \hat{\beta}_1 \bar{x} \\ \hat{\beta}_1 = \dfrac{\sum\limits_{i=1}^{n}(x_i - \bar{x})(y_i - \bar{y})}{\sum\limits_{i=1}^{n}(x_i - \bar{x})^2} \end{cases} \tag{11}$$

where $\bar{x} = \sum\limits_{i=1}^{n} x_i$, $\bar{y} = \sum\limits_{i=1}^{n} y_i$. Thus we obtain the regression equation

$$y = \hat{\beta}_0 + \hat{\beta}_1 x \tag{12}$$

With these three dynamic consumption processes, we get water demand sequence which changes over time and is influenced by population growth and economic development as follows:

$$D_t = D_{at} + D_{it} + U_{dt} \cdot P_{dt} \tag{13}$$

3.2   The Supply Model

We divide the water supply into two parts: clean water from natural source which inherently depends on the geological, topographical and ecological conditions, and fresh water from regeneration (different industries have different renewable rate). Plus, increasingly serious climate growth and pollution reduce the water provision.

The key equation is

$$S_t = \left(N \cdot A_t + SE_t + \sum_{kt} R_k \cdot D_{kt}\right)(1 - p_t) \tag{14}$$

where S represents the total supply quantitation, N represents natural water source, A represents access to clean and fresh water (rely on techniques, management and infrastructures), SE represents the water from desalination plants, R represents renewable rate of water, D represents water needs, p represents the degree of environmental pollution, and $k = a, i, d$ represents agriculture, industry and domestic, respectively.

3.3   The Demand-Supply Model

In conclusion, we define ρ as an indicator which shows the relative situation of water supply and water demand. In other words, it is a measure of the ability of a region to provide clean water to meet the needs of its population, and it contains all the dynamic factors as mentioned above.

$$\rho = \frac{S_t}{D_t} = \frac{(N \cdot A_t + SE_t + \sum\limits_{kt} R_k \cdot D_{kt})(1 - p_t)}{D_{at} + D_{it} + U_{dt} \cdot P_{dt}} \tag{15}$$

When the value of ρ is in the following interval, the ability to provide clean water to meet the needs and corresponding degree of water scarcity are shown in Table 2.

Table 2. The results of traffic flow prediction with 15-min time interval

The degree of water scarcity	The value of ρ
Very poor ability and severe water scarcity	≤ 0.7
Low ability and serious water scarcity	0.7–1
Medium ability and little water scarcity	1–1.2
High ability and no water scarcity	≥ 1.2

3.4   Evaluation

In this section, we evaluate the two pivotal models both in theory and in practice to prove its stability and feasibility.

3.4.1   Evaluating the Grey Verhulst Prediction Model

Making $\varepsilon(k)$ represents relative error, we then calculate

$$\varepsilon(k) = \frac{x^0(k) - \hat{x}^0(k)}{x^0(k)}, k = 1, 2, ..., n \tag{16}$$

where $\hat{x}^0(1) = x^0(1)$. If $\varepsilon(k) < 0.5$, we consider that it meets the general requirements; and if $\varepsilon(k) < 0.2$, we consider that it meets a high requirement.

We choose Romania and Cyprus as the test objects. Their agricultural and industrial water consumptions from the year 1982 to 2012 are shown in Table 3. We can see that only two relative errors are beyond a reasonable interval, that is, our model possesses a high accuracy.

Table 3. Real/predicted water needs of agriculture/industry in Romania and Cyprus

Year	Real needs	Predicted needs	Residuals	Relative error	Year	Real needs	Predicted needs	Residuals	Relative error
Romania_Agricutrue					*Romania_Industry*				
1982	36.12	36.12	0	0	1982	52.18	52.18	0	0
1987	41.41	45.156	−3.746	0.090	1987	45.56	47.702	−2.142	0.047
1992	44.59	34.337	10.252	0.229	1992	44.39	51.234	−6.844	0.154
1997	23.92	26.110	−2.190	0.091	1997	59.63	55.027	4.602	0.077
2002	12.93	19.854	−6.924	**0.535**	2002	66.91	59.100	7.809	0.116
2007	13.04	15.097	−2.057	0.157	2007	66.91	63.475	3.434	0.051
2012	17.03	11.480	5.549	0.325	2012	61.08	68.174	−7.094	0.116
Cyprus_Agricutrue					*Cyprus_Industry*				
1982	79.1	79.1	0	0	1982	1.83	1.83	0	0
1987	77.5	73.644	3.855	0.049	1987	1.94	1.805	0.134	0.069
1992	73.45	74.618	−1.168	0.015	1992	2.212	2.043	0.168	0.076
1997	73.93	75.605	−1.675	0.022	1997	2.37	2.311	0.058	0.024
2002	76.48	76.604	−0.124	0.001	2002	1.663	2.615	−0.952	**0.572**
2007	68.98	77.617	−8.637	0.125	2007	3.704	2.958	0.745	0.201
2012	86.41	78.643	7.766	0.089	2012	3.261	3.347	−0.086	0.026

3.4.2   Evaluating the Regression Model

This is a significant test where we take $H_0 : \beta = 0$ as the null hypothesis and $H_1 : \beta \neq 0$ as the alternative hypothesis. We construct a test statistic as follows:

$$T = \frac{\hat{\beta}_1}{\hat{\sigma}} \sqrt{\sum_{i=1}^{n} (x_i - \bar{x})^2} \tag{17}$$

where $\hat{\sigma} = \frac{1}{n-2} \sum_{i=1}^{n} \left(y_i - \hat{\beta}_0 - \hat{\beta}_1 x_i \right)^2$

When H_0 is met, T belongs to the distribution $t(n-2)$. Thus at the α confidence level, the rejection region is $\{ T \geq t_{\alpha/2}(n-2) \}$.

4   The Advanced Water Poverty Index (AWPI)

In this section, we improve the water poverty index by adding population as the sixth key factor, for population growth rate has already continually exacerbated the severe water scarcity problem [2–4,6,10,11].

We adopt the Advanced Water Poverty Index (AWPI) to measure the water supply of a country or a region, where we take the six components (Table 4) into account.

For every index, we decompose it into several sub-indexes and calculate it as below:

Table 4. The key components of the AWPI and their definitions

Components	Definitions
Resource (R)	Physical amount and distribution of internal river flows and groundwater from rainfall
Access (A)	Availability to natural water source
Capacity (C)	Ability to get clean fresh water
Use (U)	The comprehensive efficiency of water uses for agricultural, industrial and residential purposes
Environment (E)	Stability of the whole ecosystem
Population (P)	Population growth rate and population density

- Resource (The World Bank, 2015):

$$Resource\ Index\ (RI) = renewable\ internal\ freshwater\ resources\ index \tag{18}$$

- Access:

$$Access\ Index\ (AI) = \frac{S + W}{2} \tag{19}$$

where S is the sanitation coverage index (The World Bank, 2014), W is the percentage of the population with access to improved water sources, urban (The World Bank, 2015).
- Capacity:

$$Capacity\ Index\ (CI) = \frac{M + T_c + R + E + (1 - A)}{4} \tag{20}$$

where M is the evaluation of management, T_c is the evaluation of technology advances, R is the recycle rate of used water, A is the average death age index.

$$
\begin{aligned}
E &= 1, if\ average\ schooling\ year \geq 12 \\
&= 0.8, if\ 9 < averag\ schooling\ year < 12 \\
&= 0.6, if\ 6 < average\ schooling\ year < 9 \\
&= 0.4, if\ 3 < average\ schooling\ year < 6 \\
&= 0.2, if\ 1 < average\ schooling\ year < 3 \\
&= 0, if\ average\ schooling\ year < 1
\end{aligned}
\tag{21}
$$

- Use:

$$Use\ Index\ (UI) = \frac{U - U_a - U_i - U_d}{3} \tag{22}$$

where $U_k = \frac{Actual\ Water\ Consumptions\ in\ k\ Industry}{Water\ Consumption\ in\ k\ Industry}$ and (k = a,i,d) represents agriculture, industry and domestic, respectively.

- Environment:

$$Environmental\ Index\ (EI) = \frac{(1-C) + G + T + (1-P_m) + (1-P_s) + F + H}{7}$$

(23)

where C is the evaluation of influence caused by climate change, G is geological condition, T is topographical feature, P_m is average exceedance of particulate matter, P_s is average exceedance of sewage, F is the evaluation of forest health, H is the evaluation of habitat and biodiversity health.
- Population:

$$Population\ Index\ (PI) = 1 - \frac{P_g + P_d}{2}$$

(24)

where P_g is the population growth rate, P_d is the population density.
In conclusion:

$$AWPI = RI + AI + CI + UI + EI + PI$$

(25)

5 Sensitivity

When we develop the demand model to predict agricultural and industrial demand for water, we use the data from the World Bank as the initial data to simulate the trend. However, all into the long time span of the data, it's difficult for the World Bank to investigate all developing countries one by one. Thus there may exist some inaccurate initial data, which will impact the accuracy of the prediction model. We call these inaccurate initial data "Outlier".

When testing our model, we find an obvious bias in the prediction of demand in 2002. Thus, we delete the data in 2002 and use the remaining data to re-predict the demand by our model. Take Romania as an example, from Table 5, it is clear that our model is not sensitive to outliers.

Table 5. Agricultural and industrial demand with/without outliers in Romania

Year	Agricultural demand			Industrial demand		
	With outliers	Without outliers	Change rate	With outliers	Without outliers	Change rate
1982	36.12	36.12	0.00%	52.18	52.18	0.00%
1987	45.157	46.019	1.90%	47.703	44.596	−6.50%
1992	34.337	34.595	0.70%	51.234	49.94	−2.50%
1997	26.11	26.006	−0.40%	55.027	55.923	1.60%
2002	19.854	19.55	−1.50%	59.101	62.624	6.00%

6   Conclusion

The key point to measure the ability of a region to satisfy its water consumption is whether the supply can meet the demand. To measure water supply capacity of a country, it necessitates consideration of the supply and demand situation respectively. In our paper, we separate the measure model into two parts. First, we incorporate the Grey Verhulst model, the Logistic Growth and the Regression model in the demand model. Then, we develop the supply model and take various factors into account.

Moreover, we adopt the Advanced Water Poverty Index (AWPI) to measure the water supply and analyze the reasons of water scarcity of a country or a region from another angle, where we consider six key components and define each one clearly. The experimental results on World Bank dataset have proved the stability and feasibility of our models. According to the evaluation results, all the countries can take some specific intervention measures to help alleviate the severe water shortage and achieve sustainable development of water resources.

References

1. Cui, J., Liu, S.F., Zeng, B., Xie, N.M.: Parameters characteristics of grey Verhulst prediction model under multiple transformation. Kongzhi Yu Juece/Control Decis. **28**(4), 605–608 (2013)
2. Feitelson, E., Chenoweth, J.: Water poverty: towards a meaningful indicator. Water Policy **4**(3), 263–281 (2002)
3. Garriga, R.G., Foguet, A.P.: Improved method to calculate a water poverty index at local scale. J. Environ. Eng. **136**(11), 1287–1298 (2010)
4. Gong, L., Jin, C.: Urban water security evaluation system based on water poverty index. Shuili Fadian Xuebao/J. Hydroelectr. Eng. **33**(6), 84–90 (2014)
5. Kahil, M.T., Diner, A., Albiac, J.: Modeling water scarcity and droughts for policy adaptation to climate change in arid and semiarid regions. J. Hydrol. **522**, 95–109 (2015)
6. López Álvarez, B., Ramos Leal, J.A.: Water poverty index in subtropical zones: the case of Huasteca Potosina, Mexico. Rev. Int. Contam. Ambient. **31**(2), 173–184 (2015)
7. Mekonnen, M.M., Hoekstra, A.Y.: Four billion people facing severe water scarcity. Sci. Adv. **2**(2), e1500323 (2016)
8. Postel, S.L.: Entering an era of water scarcity: the challenges ahead. Ecol. Appl. **10**(10), 941–948 (2008)
9. Santini, G.: Coping with water scarcity. Unesco Tech. Doc. Hydrol. **58**(4), 77–98 (2015)
10. Sullivan, C.: Calculating a water poverty index. World Dev. **30**(7), 1195–1210 (2002)
11. Sullivan, C., Meigh, J., Lawrence, P.: Application of the water poverty index at different scales a cautionary tale. Water Int. **31**(3), 412–426 (2006)
12. Tsoularis, A., Wallace, J.: Analysis of logistic growth models. Math. Biosci. **179**(1), 21–55 (2002)
13. Wang, Z.X., Dang, Y.G., Liu, S.F.: Unbiased grey Verhulst model and its application. Syst. Eng.-Theory. **29**(10), 138–144 (2009)

Efficient Stance Detection with Latent Feature

Xiaofei Xu[1], Fei Hu[1,2], Peiwen Du[1], Jingyuan Wang[1], and Li Li[1(✉)]

[1] School of Computer and Information Science, Southwest University,
Chongqing 400715, China
xxfwin@vip.qq.com, lily@swu.edu.cn
[2] Chongqing University of Education, Chongqing 400065, China

Abstract. Social platforms, such as Twitter, are becoming more and more popular. However it is hard to identify the sentimental stance from those social media. In this paper, an approach is proposed to identify the stance of opinion. Digging out the latent factors of the given rough processed information is essential because it has the potential to reveal different aspects of the known information, which eventually contributes to the advancement of stance analysis. Generally, we take a very large number of articles from Chinese wikipedia as the corpus. The latent feature vectors are generated by word2vec. The HowNet sentiment dictionary (with positive and negative words) are applied to divide the items in the corpus into two parts. The two parts with sentiment polarity are used as the training set for SVM model. Experimentation on NLPCC 2016 Stance Detection dataset demonstrates that the proposed approach can outperform the baselines by about 10% in the term of precision.

Keywords: Word2vec · Stance detection · SVM · HowNet

1 Introduction

With the rapid growth of social platforms and mobile devices, millions of people are more likely to share their attitudes or opinions online in a timely fashion. Understanding the stance of their opinions can be very helpful in many applications [1]. For example, online reviews have greater economic impact on both consumers and companies compared with traditional media. Stance detection on news comments can be helpful for governments to improve the quality of their service. The stance analysis of product reviews is valuable for recommendation, election poll and crisis management [2]. It may uncover what really happened and what will happen by answering questions such as what is consumer's confidence on the product? What do people think about the candidates (of political parties)? What are the market trends of a particular product? These are just a few examples illustrating why stance analysis has become such a hot research topic.

Much effort has been dedicated to exploring sentiment classification via online generated texts. Different classifiers have been trained to predict the unlabeled data [3].

© Springer International Publishing AG 2017
S. Song et al. (Eds.): APWeb-WAIM 2017 Workshops, LNCS 10612, pp. 21–30, 2017.
https://doi.org/10.1007/978-3-319-69781-9_3

Although much classification work has been done, previous efforts only treated each word as an atomic unit and not all of them considered the similarity and relations of the words [4]. Those techniques have significant limitations because neither syntagmatic nor latent attributes of the contents has been targeted properly.

This may lead to a delusive faith in stance detection tasks. Inspired by traditional sentiment classification methods, we combine SVM with word embedding [5] to process stance detection at word level. The HowNet sentiment dictionary is used as the basis. Chinese wikipedia text data is used as the corpus. We further process the stance detection at document level using the stance words generated by SVM.

The rest of the paper is organized as follows. In Sect. 2, we briefly review related work. Section 3 mainly presents the architecture of our method. Section 4 illustrates the experimental results, followed by the discussions. Section 5 concludes this paper.

2   Related Work

A great deal of work has been done on the problem of sentiment classification at different levels, ranging from document level, sentence level to word level. The main idea is to identify the users' sentiment inclination as either positive, negative or neutral from the obtained information. Turney and Littman [6] used an unsupervised learning algorithm with mutual information to predict the semantic orientation at word level. Pang and Lee [7] proposed a semi-supervised machine learning algorithm by employing a subjectivity detector at sentence level. A generative model that jointly models sentiment words, topic words and sentiment polarity in a sentence as a triple was proposed by Eguchi and Lavrenko [8] At the document level. From earlier experience, we argue that the better word representation is crucial to the success of stance detection. The paper tries to capture the latent features of the words from a large corpus with the distributed representation proposed in [9].

2.1   Word Embedding

Models based on Bag of Words (BOW) [10] have been proved promising in Natural Language Processing (NLP). They put repeated words superimposed to increase confidence of word in a sentence. In BOW models, sentences are mapped into a vector space that each dimension represents a unique word. The more times a word repeats in a sentence, the higher the weight of the dimension of the word is. BOW assumed that every word is independent of each other. It is the one-hot representation which suffered from the inherent limitation. Then many improvements on BOW are proposed, such as the Locally Weighted BOW (LOWBOW) [11] and semantic hashing representation [12].

The idea mapping each word into a vector space was first proposed by Hinton [13], which determined the semantics similarity of two words by calculating the distance of them in the vector space, such as the cosine distance, the Euclidean distance or the Hamming distance. After Bengio *et al.* [14] proposed the neural language model, the study of the word embedding technology had been growing gradually. Based on the speech tagging and command-entity recognition technologies, Collobert *et al.* [15] obtained words embedding by semi-supervised training in corpus. Huang *et al.* [16] extracted richer semantic information from the training set, unlike other models only focusing on the context of limited recent words, they were interested in extracting semantic information from the whole document using a BOW model. Mikolov *et al.* [17] developed a tool named word2vec for word embedding.

The word2vec[1] includes two models: the CBOW (Continuous Bag-Of-Word) model and the Skip-gram (continuous Skip-gram) model. Both models are the feed-forward neural network models, without the nonlinear hidden layer of the neural language model.

2.2   SVM Classification

As one of the discriminative classification methods, Support Vector Machine (SVM) classification has been shown to be more accurate than other classification approaches. Ye *et al.* [18] takes advantage of SVM to classify sentiment of review, but it doesn't consider the sentiment of word. Gayathri and Marimuthu [19] proposes a text classification based on the feature selection by reducing the dimensionality of the feature vector. Dave *et al.* [20] studies the sentiment classification as well but focusing on several product reviews included MP3, PDA, TVs, laptop etc. from Amazon. Inspired by the existing work, we propose a classifier by combining SVM with the word vector to predict the sentiment inclination at the document level. The HowNet is employed to expand the sentiment dictionary. More details will be discussed subsequently.

3   Latent SVM

In this section, we will introduce latent SVM, the proposed method for stance detection.

A massive collection of corpus is used as the word embedding data. After conducting word segmentation and removing the stop-words, a huge word list is obtained. The distributed vector representation of each word is generated by word2vec. The size of the vector has an important impact on the final result. We will have a detailed discussion later.

Then we use HowNet and SVM to build a classification model based on one group of the word list. The model is then used to predict the sentiment orientation of the remaining words from another group of the list. For simplicity, the task here can be thought of as binary classification problem. Currently, the sentiment stance of the whole document is decided by the majority rule, one

---

[1] https://code.google.com/archive/p/word2vec/.

of the most influential and widely used binary decision rules. In other words, by comparing the number of positive words with the number of negative words in the document separately, we will acquire the sentiment orientation of the document.

More concretely, suppose there are q documents, we use $(w_{j1}, w_{j2}, \ldots, w_{jp_j})$ to represent document j with p_j words, S the document-word matrix. We then obtain m distinct words from S, called *terms* in the paper.

$$terms = \{t_1, t_2, ..., t_m\} \tag{1}$$

$$S = \begin{pmatrix} w_{11}, w_{12}, w_{13}, \ldots, w_{1p_1} \\ \vdots \\ w_{q1}, w_{q2}, w_{q3}, \ldots, w_{qp_q} \end{pmatrix} \tag{2}$$

Then with S, the input of word2vec algorithm introduced in Sect. 2.1, we get the vector set of S as matrix V, each term t_c is presented as n-dimensional vector such as $(v_{c1}, v_{c2}, \ldots, v_{cn})$. V is notated as follows:

$$V = \begin{pmatrix} v_{11}, & v_{12}, & v_{13}, & \ldots, & v_{1n} \\ & & \vdots & & \\ v_{m1}, & v_{m2}, & v_{m3}, & \ldots, & v_{mn} \end{pmatrix} \tag{3}$$

(1) Polarity annotation

With the obtained matrix V, we then consult the HowNet sentiment dictionary to pick up all words in the dictionary from V. The polarity label, either positive (marked by $+1$) or negative (marked by -1), is then certain for each selected word. The annotated vector representations are confederated as the training set.

(2) Model training

SVM model is the main classifier in our work. To optimize SVM model result, we use a grid traversal algorithm to get the near-optimal solutions of the *cost* and γ parameters.

(3) Dictionary expansion

Once the training process completed, the classifier can act as the predefined template for similar resource classification. Recall that we grouped the big corpus into two parts during the annotation phrase. The unlabeled part, with exactly the same word embedding, can be replaced because the massive Chinese wikipedia corpus, which is broad in scope and content, is able to add deep semantics to the task processing. It is generally presumed that a comprehensive sentiment dictionary is crucial to the sentiment analysis. So we attempt to label the remaining part of the corpus with the trained classifier and eventually expand the sentiment dictionary. Finally we use voting algorithm to judge the stance of each document.

Similar pattern within the input vectors highly reflects the relevance of words. The trained classifier can detect the pattern by providing informative clues on why it means that. This in turn provides insight into the content itself and inspires our further discussion in Sect. 4.

4 Experiments and Results

4.1 Preprocessing

Dataset. In this work, we choose a dump of Chinese wikipedia data, including Arts, History, Society, Technology, Entertainment and some other categories. We select the most recent versions of all articles from 2004 to May 15, 2015[2]. During preprocessing phrase *xml* labels are transformed into the plain texts. We then remove all *html* labels and punctuations. Finally we get 8,676,665 documents and 205, 427, 273 words in total.

The word segmentation tool jieba[3] is employed for Chinese word segmentation (including the compound words).

To show the performance of our approach on stance detection tasks and traditional sentiment classification tasks, we choose two testing sets. The Sentiment Classification of Chinese Microblogs data released by NLPCC 2014[4] and Stance Detection of Chinese Microblogs data at NLPCC 2016[5]. Two datasets provide raw texts in Chinese Microblogs with their stance annotations. The testing dataset for NLPCC 2014 is formed by 5,000 pieces of positive data and 5,000 pieces of negative data, respectively, while in NLPCC 2016 data, all tagged data is used. Following the same process, firstly segmenting the words, then removing stop-words and finally two testing datasets are obtained.

Evaluation tasks and metrics. For training data and validate data, we use 10-fold cross-validation on each dataset to reduce errors. Here we use a series of baselines for comparison. Precision, Recall and F1-Score are the evaluation metrics.

4.2 Baselines

The methods used in the experiments are as follows.

SVM: It is a simple and efficient classification method that can use vector as input.

NB and BN: We use Naive Bayes and Bayesian Network as training method, and takes the same features mentioned in evaluation tasks.

C4.5: It is an improved series of algorithm used in the classification of machine learning.

BP: It is a widely used algorithm in neural network training.

Besides the classification methods mentioned above, we use Lantent Semantic Analysis (LSA) [21] as the contrast of the word vector training methods.

---

[2] http://dumps.wikimedia.org/zhwiki/20150515.

[3] http://github.com/fxsjy/jieba.

[4] http://tcci.ccf.org.cn/conference/2014/dldoc/evans01.zip.

[5] http://tcci.ccf.org.cn/conference/2016/dldoc/evasampledata4-TaskA.txt.

4.3   Experimental Results

In our experiment, we use Chinese wiki data mentioned above to build our expanded dictionary. Then, use these dictionaries to perform stance detection on our test data. Tables 1 and 2 list the performance of different methods on NLPCC 2014 and NLPCC 2016 dataset.

Table 1. NLPCC 2014: performance of classification

Embedding	Method	Precision	Recall	F1-Score
LSA	BP	0.567	0.543	0.555
	NB	0.719	0.507	0.595
	BN	0.746	0.503	0.601
	C4.5	0.69	0.543	0.608
	SVM	0.121	0.629	0.203
Word2vec	BP	0.724	0.593	0.652
	NB	0.727	**0.634**	**0.677**
	BN	0.719	0.624	0.668
	C4.5	0.682	0.594	0.635
	LatentSVM	**0.821**	0.574	0.676

Table 2. NLPCC 2016: performance of classification

Embedding	Method	Precision	Recall	F1-Score
LSA	BP	0.537	0.521	0.529
	NB	0.679	0.497	0.574
	BN	0.776	0.492	0.602
	C4.5	0.627	0.491	0.551
	SVM	0.201	**0.779**	0.32
Word2vec	BP	0.644	0.557	0.597
	NB	0.725	0.625	0.671
	BN	0.734	0.635	0.681
	C4.5	0.67	0.595	0.63
	LatentSVM	**0.883**	0.583	**0.702**

Out method achieves better performance than the baselines. In terms of precision, our method achieved about 10% improvement on average in NLPCC 2016 data set. However our method performs not well in recall as illustrated in Tables 1 and 2. The NB algorithm achieves the best recall and F1-score in

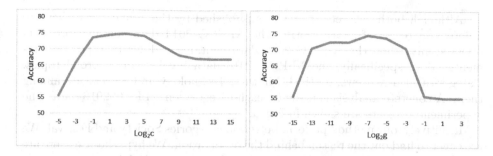

Fig. 1. SVM cost parameter & accuracy   **Fig. 2.** SVM gamma parameter & accuracy

NLPCC 2014 data, while the SVM with LSA having the best recall in NLPCC 2016 data. Actually, SVM with LSA performs very bad in term of precision. Our method can acquire the best in both big and small dataset in term of precision.

In our experiment, we use a series of parameters of SVM to optimize the performance of our classification model, actually this optimize method is effective. As is shown in Figs. 1 and 2, we got best accuracy of 88.7% after optimized SVM model. Then, we turn to word2vec parameter for further improvement. We test series of data to ensure these parameter are effective. As is shown in Figs. 3 and 4, we test word2vec vector dimension size with our data and find that we can gain the best value by changing the parameter of vector dimension size.

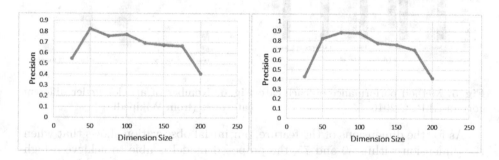

Fig. 3. NLPCC 2014 dimension size & pre-   **Fig. 4.** NLPCC 2016 dimension size cision   & precision

4.4 Discussions

The whole process is suggestive of new perspectives. We are interested in the role the distributed word representation plays in our work. Is there any contribution from the massive corpus to the final result? Or what does the large corpus mean to the stance detection? In the following, we will try to answer these questions and provide some hints on how to use the above information. First, we focus on the testing data.

Although both datasets are reviews published by the online users, they are different in terms of subcategories. the NLPCC 2014 data contains products reviews only while the NLPCC 2016 data contains 5 different types of comments. More specifically, the NLPCC 2016 data involves more broad topics. Categories such as News, Festival, Military, Technology and Society are among them. A concrete analysis is made on each category on NLPCC 2016 data. The experimental results are shown in Fig. 5.

Intuitively, our method perform better on categories Society and Festival. We are eager to find out the reason behind the observation. We hypothecate there are two explanations for that. One is the training corpus itself. We are speculating whether a rather large corpus works for our task in the paper. The other is the dimensions of the word embedding representation. We first turn to Wikipedia official website for more information about the detailed statistics on the number of articles Wikipedia includes so far. We are only interested in the categories included in NLPCC 2016 dataset. The result is shown in Fig. 6. Analyzing the results shown in Figs. 5 and 6, we can not see any necessary consequence between two figures. Perhaps after word segmentation over the wiki articles, some inherent relations between/among those words are disconnected.

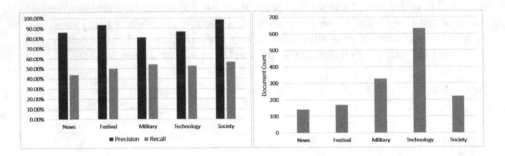

Fig. 5. Method performance on each category in NLPCC 2016   **Fig. 6.** Number of articles under different categories (from Wikipedia)

As for the dimensions of the feature, our initial observation shows that when the dimension within 50 and 75, the proposed model is able to achieve a high performance. To further discover the underlying rationale, we determine to test the model again with a step 5 at each run. Two categories (Military and Society) are deliberately selected for testing purpose. We also want to further demonstrate whether the correlation between the corpus and the final result exists or not. The results are shown in Fig. 7. It shows that the dimension of the feature, do influence the overall performance. For example, when the dimension is 50 and 75, respectively in Fig. 7, the precision can rise up to 99.9%. In other words, 50 and 75 are two influential points for category Society.

In short, it is clear that the dimension has great impact on the performance of the model. The proposed model works better on some categories. We expect further analyzing the distribution of the words in the corpus and each category as well to give more hints on the improvement of the model.

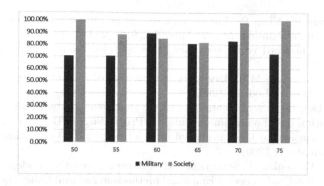

Fig. 7. Different feature sizes on military and society topic

5   Conclusion

In this paper, we introduce and analyze a document stance detection method which combines latent features with the SVM classifier. The distribution representation is able to catch the deep stance of the word. Based on the popular word2vec algorithm, a large corpus consisting of a large number of Chinese wiki articles is used to generate word embedding representation for each word. SVM classifier is applied to expand the stance word dictionary. Currently, the stance of the whole document is decided by the majority rule. The experiment results on two datasets demonstrates that the proposed approach can outperform the baselines about 10% in terms of precision.

There are also some interesting findings which merit further studying. The HowNet-based stance detection is very useful, however, there is a serious lack of hotspot information. Hotspot information detection and analysis is prevalent in social networks and also in stance detection. We are planning to investigate the underlying reasons why our model does not achieve higher performance in terms of recall metric when compared with the baseline models. Additionally, it will be interesting to expand the stance dictionary further to collect an even broader vocabulary.

Acknowledgement. This work was supported by National Undergraduate Training Program for Innovation and Entrepreneurship (No. 201610635004).

References

1. Pang, B., Lee, L.: Opinion mining and sentiment analysis. Found. Trends Inf. Retr. **2**(1–2), 1–135 (2008)
2. Hirschberg, J., Manning, C.D.: Advances in natural language processing. Science **349**(6245), 261–266 (2015)
3. Go, A., Bhayani, R., Huang, L.: Twitter sentiment classification using distant supervision. CS224N Project report, Stanford, vol. 1, p. 12 (2009)

4. Koteeswaran, S., Visu, P., Janet, J.: A review on clustering and outlier analysis techniques in datamining. Am. J. Appl. Sci. **9**(2), 254 (2012)
5. Wurst, M.: The word vector tool user guide operator reference developer tutorial (2007)
6. Turney, P., Littman, M.L.: Unsupervised learning of semantic orientation from a hundred-billion-word corpus (2002)
7. Pang, B., Lee, L.: A sentimental education: sentiment analysis using subjectivity summarization based on minimum cuts. In: Proceedings of the 42nd Annual Meeting on Association for Computational Linguistics, p. 271. Association for Computational Linguistics (2004)
8. Eguchi, K., Lavrenko, V.: Sentiment retrieval using generative models. In: Proceedings of the 2006 Conference on Empirical Methods in Natural Language Processing, pp. 345–354. Association for Computational Linguistics (2006)
9. Mikolov, T., Chen, K., Corrado, G., Dean, J.: Efficient estimation of word representations in vector space. arXiv preprint arXiv:1301.3781 (2013)
10. Zhang, Y., Jin, R., Zhou, Z.-H.: Understanding bag-of-words model: a statistical framework. Int. J. Mach. Learn. Cybernet. **1**(1–4), 43–52 (2010)
11. Lebanon, G., Mao, Y., Dillon, J.: The locally weighted bag of words framework for document representation. J. Mach. Learn. Res. **8**(Oct), 2405–2441 (2007)
12. Shen, Y., He, X., Gao, J., Deng, L., Mesnil, G.: Learning semantic representations using convolutional neural networks for web search. In: Proceedings of the 23rd International Conference on World Wide Web, pp. 373–374. ACM (2014)
13. Hinton, G.E.: Learning distributed representations of concepts. In: Proceedings of the Eighth Annual Conference of the Cognitive Science Society, vol. 1, p. 12, Amherst, MA (1986)
14. Bengio, Y., Ducharme, R., Vincent, P., Jauvin, C.: A neural probabilistic language model. J. Mach. Learn. Res. **3**(Feb), 1137–1155 (2003)
15. Collobert, R., Weston, J., Bottou, L., Karlen, M., Kavukcuoglu, K., Kuksa, P.: Natural language processing (almost) from scratch. J. Mach. Learn. Res. **12**(1), 2493–2537 (2011)
16. Huang, E.H., Socher, R., Manning, C.D., Ng, A.Y.: Improving word representations via global context and multiple word prototypes. In: Proceedings of the 50th Annual Meeting of the Association for Computational Linguistics: Long Papers, vol. 1, pp. 873–882. Association for Computational Linguistics (2012)
17. Mikolov, T., Deoras, A., Kombrink, S., Burget, L., Cernockỳ, J.: Empirical evaluation and combination of advanced language modeling techniques. INTERSPEECH, vol. 1, pp. 605–608 (2011)
18. Ye, Q., Lin, B., Li, Y.-J.: Sentiment classification for Chinese reviews: a comparison between SVM and semantic approaches. In: 2005 International Conference on Machine Learning and Cybernetics, vol. 4, pp. 2341–2346. IEEE (2005)
19. Gayathri, K., Marimuthu, A.: Text document pre-processing with the KNN for classification using the SVM. In: 2013 7th International Conference on Intelligent Systems and Control (ISCO), pp. 453–457. IEEE (2013)
20. Dave, K., Lawrence, S., Pennock, D.M.: Mining the peanut gallery: opinion extraction and semantic classification of product reviews. In: Proceedings of the 12th International Conference on World Wide Web, pp. 519–528. ACM (2003)
21. Landauer, T.K., Foltz, P.W., Laham, D.: An introduction to latent semantic analysis. Discourse Process. **25**(2–3), 259–284 (1998)

Deep Analysis and Utilization of Malware's Social Relation Network for Its Detection

Shifu Hou[1], Lingwei Chen[1], Yanfang Ye[1(✉)], and Lifei Chen[2]

[1] Department of Computer Science and Electrical Engineering,
West Virginia University, Morgantown, WV 26506, USA
{shhou,lgchen}@mix.wvu.edu, yanfang.ye@mail.wvu.edu
[2] School of Mathematics and Computer Science, Fujian Normal University,
Fuzhou 350117, Fujian, China
clfei@fjnu.edu.cn

Abstract. To combat with the evolving malware attacks, many research efforts have been conducted on developing intelligent malware detection systems. In most of the existing systems, resting on the analysis of file contents extracted from the file samples (e.g., binary n-grams, system calls), data mining techniques such as classification and clustering have been used for malware detection. However, ignoring the social relations among these file samples (i.e., utilizing file contents only) is a significant limitation of these malware detection methods. In this paper, (1) instead of using file contents extracted from the collected samples, we conduct deep analysis of the social relation network among file samples and study how it can be used for malware detection; (2) resting on the constructed file relation graph, we perform large scale inference by propagating information from the labeled samples (either benign or malicious) to detect newly unknown malware. A comprehensive experimental study on a large collection of file sample relations obtained from Comodo Cloud Security Center is performed to compare various malware detection approaches. Promising experimental results demonstrate that the accuracy and efficiency of our proposed method outperform other alternate data mining based detection techniques.

Keywords: Malware detection · Social relation network · Graph inference

1 Introduction

Nowadays, as computers and Internet become increasingly ubiquitous, especially the rapid development of e-commerce, computer security becomes more and more important. Malware (short for *mal*icious soft*ware*) is software that deliberately fulfills the harmful intent of an attacker [3], such as viruses, trojans, worms and botnets. It has been used as the major weapon by the cyber-criminals to launch a wide range of security attacks which present serious damages and significant financial loss to Internet users [10]: the average infected computers per day was

S. Song et al. (Eds.): APWeb-WAIM 2017 Workshops, LNCS 10612, pp. 31–42, 2017.
https://doi.org/10.1007/978-3-319-69781-9_4

between 2–5 millions [14] and the average loss caused by malware attacks was around $345,000 dollars per incident [11]. To protect legitimate users from the attacks, the most significant line of defense against malware is anti-malware software products, such as Comodo, Kaspersky and Symantec Anti-Virus. Typically, these widely used malware detection software tools use the signature-based method [4] to recognize threats. However, driven by considerable economic benefits, malware attackers have invented automated malware development toolkits (such as Zeus [23]) to create and mutate thousands of malicious codes per day which can bypass the traditional signature-based detection. In order to effectively and automatically detect these large new generated malware samples, intelligent malware detection systems have been developed by applying data mining techniques [2,13,15,21,26,28,29]. Such techniques have successes in classifying or clustering particular sets of malware samples.

Although utilizing file contents only, either static or dynamic extractions, and simply treating the files as independent samples may allow many off-the-shelf classification or clustering tools to be directly adapted for malware detection, ignoring the social relations among file samples can be a significant limitation of current malware detection methods, since the usual *i.i.d* (independent and identical distributed) assumption may not hold for malware samples. Actually, the social relations among the file samples (e.g., whether the files co-exist in the users' computers, whether the files are created at the same time, etc.) may imply the inter-dependence among them and can provide invaluable information about their properties. For example, if a file is always associated with many trojans in users' computers, then most likely, it is a malicious Trojan-Downloader file. In this paper, instead of using file contents extracted from the collected samples, we conduct deep analysis of the social relation network among file samples and study how it can be used for malware detection. Based on the constructed file relation graph, large scale inference by propagating information from the labeled samples (either benign or malicious) is performed to detect newly unknown malware. A comprehensive experimental study on a large collection of file sample relations obtained from Comodo Cloud Security Center is performed to compare various malware detection approaches. Promising experimental results demonstrate that the accuracy and efficiency of our proposed method outperform other alternate data mining based detection techniques. The contributions of this paper can be summarized as follows:

- *Deep Analysis of Malware's Social Relation Network:* Different from file content based detection, we analyze and utilize the social relations among file samples (i.e., co-existences of the files) collected from the user clients to construct file relation graph for malware detection. The newly unknown malware can be detected by its association with the known files (benign or malicious).
- *An Improved Graph Inference Algorithm for Unknown File Labeling:* Belief Propagation (BP) algorithm is a promising method for solving inference problems over graphs and it has also been successfully used in many domains (e.g., computer vision, coding theory) [30]. However, in our application, the algorithm should be greatly adapted, which is not a trivial process: we fine tune

various components used in the algorithm and carefully design the message update and belief read-out functions for malware detection.

– A *Comprehensive Experimental Study based on Real Data Collection from Industry:* Based on the real sample set and the co-existence relationship among the files obtained from Comodo Cloud Security Center, we construct the file relation graph and provide a comprehensive experimental study to evaluate our proposed method.

The rest of the paper is organized as follows. Section 2 discusses the related work. Section 3 presents file relation graph construction method and provides a deep analysis of malware's social relation network. Section 4 introduces our proposed graph inference algorithm for malware detection based on the constructed file relation graph. In Sect. 5, using the real data collection obtained from Comodo Cloud Security Center, we systematically evaluate the effectiveness and efficiency of our proposed method in comparison with other alternate data mining approaches. Finally, Sect. 6 concludes.

2 Related Work

Signature-based method [4] is widely used in anti-malware industry for malware detection. However, malware attackers can easily evade this signature-based method through techniques such as encryption, packing, obfuscation, polymorphism, and metamorphism [5]. To gain profits, today's malware samples are created at a high speed (thousand per day). In order to remain effective, intelligent malware detection systems have been developed by applying data mining and machine learning techniques [1,2,15,16,25,26,29]. In these systems, the detection process is generally divided into two steps: *feature extraction* and *classification/clustering*. In the first step, various features, such as Windows Application Programming Interface (API) calls [29], program strings [15,20], and behavior based features [12], are extracted to capture the characteristics of the file samples. In the second step, classification or clustering techniques are used to automatically classify the file samples into different classes based on computational analysis of the feature representations. These intelligent malware detection systems are varied in their use of feature representations and classification/clustering methods. Most of such techniques simply treat the files as independent samples, however, the social relations among file samples may imply the inter-dependence among them and the usual *i.i.d* (independent and identical distributed) assumption may not hold for malware samples. As a result, ignoring the relations among file samples is a significant limitation of current malware detection methods.

Actually, besides file contents, the social relations between file samples (e.g., file co-existences, file co-operations) can provide invaluable information about the properties of file samples [27]. In recent years, limited research efforts have been conducted on file relation based malware detection. Chau et al. [7] applied a graph-based approach to infer the file reputations by analyzing file-to-machine relations. Venzhega et al. [24] built regression classifier based on file placements for malware detection. In our previous work [27], we proposed a semi-parametric classification model for combining file content and file relations together for malware detection. In this paper, we will study how the relations between file samples can be used separately for malware detection. Different from the work in [8,9,22], we provide deep analysis of malware's social relation network and improve the graph inference algorithm for newly unknown malware detection.

3   Deep Analysis of Malware's Social Relation Network

In this section, we (1) first introduce the file relation graph construction, and (2) then provide deep analysis of malware's social relation network.

3.1   File Relation Graph Construction

Based on the collected file lists from the user clients, we construct a graph to describe the social relations among file samples (i.e., co-existence relationships). Generally, two files are related if they are shared by a group of clients (or equivalently, file lists). The file relation graph is defined as $G = (V, E)$, where V is the set of file samples and E denotes the relations between file samples. Given two file samples v_i and v_j, let C_i be the set of user clients containing v_i and C_j be the set of user clients containing v_j. $|.|$ represents the size of a set. We define the *strength* of the relations (i.e., co-existence) between v_i and v_j based on the overlap between sets C_i and C_j, and use Jaccard similarity for measurement

$$JS(v_i, v_j) = \frac{|C_i \bigcap C_j|}{|C_i \bigcup C_j|}. \tag{1}$$

If the strength $JS(v_i, v_j)$ between a pair of nodes (file samples) is greater than the specified threshold δ_{JS}, which indicates a *strong* (not *weak*) relation between v_i and v_j, then there is an edge between them. Each file is in a state of $S \in \{s_m, s_b, s_g\}$ (s_m: malicious, s_b: benign, s_g: unknown). The weight of edge between v_i and v_j, which is the probability of node v_i being in the state s_i and node v_j being in the state s_j, is defined as

$$w(v_i, v_j) = \frac{|E_{s_i,s_j}|}{|E|}, \tag{2}$$

where $|E_{s_i,s_j}|$ is the number of the edges between all the files with states s_i and s_j, and $|E|$ is the number of all the edges. The weight of node v_i which denotes its popularity can be defined as

$$w(v_i) = \frac{|C_i|}{|C|}, \tag{3}$$

where C is the set of all the user clients.

3.2 Graph Property Analysis

To analyze the property of malware's social relation network (co-existence relationship), we obtained a real dataset from Comodo Cloud Security Center: the dataset includes the file lists from 1,000 clients which describe file co-existence relations between 1,540 malware, 7,687 benign files and 2,250 unknown files (2,018 of them are analyzed by the anti-malware experts of Comodo Security Lab, 91 of them are malware and 1,927 of them are benign files). Figure 1 shows a zoomed-in view of a part of the constructed file relation graph (malware marked in red and benign files marked in green). From Fig. 1, we can see that many of the red nodes are associated with other red nodes and form some clusters, while the green nodes are also related to other green nodes and form their clusters. The nodes within the same cluster have strong relations with each other: (1) the red clusters may be the variants of malware families (e.g., family of online-game trojans); (2) the green clusters may be the related files of same applications (e.g., Acrobat installation archive and its related files).

Fig. 1. Visualization of a social relation network among file samples (Color figure online)

Based on the dataset described above, we also further use fourteen measures in Table 1 to see the differences between benign file relation graph, ordinary malware (i.e., 1,220 malware whose the existence frequency is < 100) file relation graph and popular malware (i.e., 320 malware whose existence frequency is ≥ 100) file relation graph. In Table 1, from the comparisons of G1 and G2, we can see that the measures of *components, component ratio, connectedness* and *fragmentation* are different between benign file relation graph and ordinary malware file relation graph; while from the comparisons of G2 and G3, we can see that the measures of *avg degree, centralization* and *density* are different between ordinary malware file relation graph and top popular malware file relation graph. The different properties between benign file relation graph and malware file relation graph enable us to discriminate malware and benign files, while the different properties between ordinary malware file relation graph and popular malware file relation graph may allow us to predict the trend of malware prevalence.

Table 1. Graph property comparisons

NO.	Measures	G1	G2	G3
1	H-Index	129	125	125
2	Avg degree	49.842	*40.677*	*12.098*
3	Centralization	0.340	*0.344*	*0.081*
4	Density	0.018	*0.014*	*0.001*
5	Components	**103**	**11**	2
6	Component ratio	**0.036**	**0.004**	0.000
7	Connectedness	**0.964**	**0.996**	1.000
8	Fragmentation	**0.036**	**0.004**	0.000
9	Closure	0.081	0.091	0.047
10	Avg distance	2.744	3.056	3.408
11	SD distance	0.631	0.836	0.717
12	Diameter	5	7	4
13	Breadth	0.625	0.645	0.689
14	Compactness	0.375	0.355	0.311

"G1": graph constructed based on 7,687 benign files and files co-exist with them, "G2": graph constructed based on 1,220 ordinary malware and files co-exist with them, "G3": graph constructed based on 320 popular malware and files co-exist with them.

4   A Graph Inference Algorithm for Unknown File Labeling

Belief Propagation (BP) algorithm is a promising method for solving inference problems over graphs and it has also been successfully used in many domains (e.g., computer vision, coding theory) [17,30]. It was first proposed by Pearl [19] to calculate marginal distribution in Markov Random Fields and Bayes Nets. Nodes of the graph perform as a local summation operation by iterations using the prior knowledge from their neighbors and then pass the information to all the neighbors in the form of messages [18]. By definition, the message is the neighbor node's opinion about the current node's probability of being in the designated status. Mathematically, the message update equation in standard BP is

$$m_{i->j}(x_j) = \sum_{x_i \in S} f_{i->j}(x_i, x_j) g_i(x_i) \prod_{k \in N(i)/j} m_{k->i}(x_i), \qquad (4)$$

where $m_{i->j}(x_j)$ is the message sent from node i to node j, that is, node i's belief that node j is in the state x_j; both $g_i(x_i)$ and $f_{i->j}(x_i, x_j)$ are typically called as energy functions, in which, $g_i(x_i)$ is the node potential, meaning the prior probability of node i being in the state x_i, while $f_{i->j}(x_i, x_j)$ is the edge potential, referring the probability of node i being in the state x_i and node j being in the state x_j; S is the set of states; $N(i)/j$ is the set of nodes neighboring node i (not including node j). BP algorithm stops when message updates converge or a maximum number of iterations has finished. Then the belief value of each node is calculated as follow

$$b_i(x_i) = k \times g_i(x_i) \prod_{k \in N(i)} m_{k->i}(x_i),$$
(5)

where k is a constant.

The standard BP is commonly called *sum-product* (from its message-update equation). A simple variant, called *max-product*, is used to estimate the state configuration with maximum probability, where the message update is the same as Eq. 4, except that sum is replaced by max, and the belief equation is the same as Eq. 5 [6].

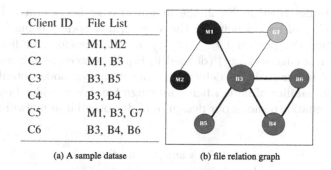

(a) A sample datase   (b) file relation graph

Fig. 2. A sample dataset and its file relation graph constructed

The standard BP has been implemented in AESOP [22] for malware detection, however, it fails in our application. To put this into perspective, we use the sample dataset in Fig. 2(a) for further illustration, in which "M" denotes malware, "B" denotes benign file, and "G" is unknown file. Figure 2(b) is the constructed file relation graph based on the sample dataset (note that the weights of the nodes and edges are different). We employ the same energy functions designed in AESOP [22]. When the message updates converge (within threshold 10^{-3}), the belief values of the data nodes (i.e., BP_Belief) are shown in Table 2. From the results (i.e., BP_Class) in Table 2, we can see that file $B3$ and file $G6$ are misclassified.

In order to solve the problem above and make BP tailor to our application, we fine tune various components in BP and carefully design the message update and belief read-out functions. Before doing that, we first analyze the meaning of each energy function in our case for malware detection. In Eq. 4, $m_{i->j}(x_j)$, $f_{i->j}(x_i, x_j)$, and $g_i(x_i)$ represent message from node i to node j, edge potential, and node potential respectively. For malware detection problem, accordingly, $m_{i->j}(x_j)$ means the probability of node i believes that the neighbor node j being a benign file; $f_{i->j}(x_i, x_j)$ is the probability that node i and node j can be connected together; and $g_i(x_i)$ is the prior probability of node i being a

Table 2. The results of standard BP and IGIA based on Fig. 2

Nodes	BP_Belief	BP_Class	IGIA_Belief	IGIA_Class
M1	0.002642	M	0.008898	M
M2	0.005500	M	0.003617	M
B3	0.000006	M	0.121844	**B**
B4	0.207126	B	0.089217	B
B5	0.633993	B	0.067515	B
B6	0.207126	B	0.010625	B
G7	0.000007	M	0.078065	**B**

benign file. As described in Eq. 2, the weight of edge between a pair of nodes is the probability of node i being in the state x_i and node j being in the state x_j, which is the edge potential $f_{i->j}(x_i, x_j)$ in BP. Therefore, we fine tune and use the weight of edge $w(x_i, x_j)$ (defined in Eq. 2) between node i and j as the edge potential in our malware detection application. For node potential, $g_i(x_i)$ is the prior probability of node i being a benign file. We consider both its state and weight. Equation 6 shows our design of node potential in malware detection problem.

$$g_i(x_i) = \begin{cases} 0.5 + 0.5 * w(x_i) & \text{if } state(x_i) = s_b \\ 0.5 & \text{if } state(x_i) = s_g \\ 0.5 - 0.5 * w(x_i) & \text{if } state(x_i) = s_m, \end{cases} \qquad (6)$$

where $w(x_i)$ is the weight of node i which can be calculated by Eq. 3 .

Instead of using sum-product, we redesign the message update equation as below

$$m_{i->j}(x_j) = \frac{1}{p} \sum_{x_i \in S} f_{i->j}(x_i, x_j) g_i(x_i) \frac{\sum_{k \in N(i)/j} m_{k->i}(x_i)}{q}, \qquad (7)$$

where q equals to the number of the neighbors of node i (excluded node j), and p is a normalizing constant. In our application, we also initialize all the messages to 1. To be more robust to outliers, the belief read-out equation is redesigned as follow

$$b_i(x_i) = \frac{1}{r} g_i(x_i) Median_{k \in N(i)} \{m_{k->i}(x_i)\}, \qquad (8)$$

where r is an adjustable constant.

The above Improved Graph Inference Algorithm is denoted as **IGIA**. Based on the energy functions as well as fine tuned message update and belief read-out equations designed in IGIA, using the same sample dataset in Fig. 2(a), the belief value of each node (i.e., IGIA_Belief) is shown in Table 2 and the results (i.e., IGIA_Class in Table 2) demonstrate that our proposed IGIA performs well in malware detection problem.

5   Experimental Results and Analysis

In this section, we conduct two set of experiments to evaluate our proposed graph inference algorithm for malware detection.

5.1   Experimental Setup

We use the same dataset obtained from Comodo Cloud Security Center as described in Sect. 3.2: the dataset includes the file lists from 1,000 clients which describe file co-existence relations between 1,540 malware, 7,687 benign files and 2,250 unknown files (2,018 of them are analyzed by the anti-malware experts of Comodo Security Lab, 91 of them are malware and 1,927 of them are benign files). The completely constructed file co-existence relation graph includes 11,477 nodes and 412,810 edges. The 1,540 malware and 7,687 benign files are used for training, while the 2,018 files labeled by human experts from the unknown sample collection are used for testing. We evaluate the malware detection performance of different methods using TP (true positive), TN (true negative), FP (false positive), FN (false negative), TPR (TP rate), FPR (FP rate), and ACY (accuracy).

5.2   Comparisons of Different Graph Inference Algorithms

In this section, we compare our proposed graph inference algorithm (IGIA) with standard BP (sum-product) implemented in AESOP [22] and BP with max-product. The results in Table 3 show that our proposed graph inference algorithm (IGIA) performs better than other two BP algorithms, due to our well designed energy functions and tuned message update as well as belief read-out.

Table 3. Comparisons of different graph inference algorithms

Training	TP	FP	TN	FN	ACY
IGIA	1,533	254	7,433	7	**0.9717**
Sum-product	1,469	7,404	283	71	0.1899
Max-product	1,244	6,142	1,545	296	0.3023
Testing	TP	FP	TN	FN	ACY
IGIA	65	55	1,872	26	**0.9598**
Sum-product	89	1,812	115	2	0.1011
Max-product	70	1,507	420	21	0.2428

5.3   Comparisons with Other Classification Approaches

In this section, we compare the malware detection effectiveness and efficiency of our proposed graph inference algorithm (IGIA) and other classification approaches (e.g., Support Vector Machine (SVM) and Decision Tree (DT)). Table 4 shows that the our proposed graph inference algorithm (IGIA) outperforms the other typical classification methods in malware detection. For malware detection efficiency, based on the 2,018 testing samples, Fig. 3 also demonstrates that the proposed graph inference algorithm (IGIA) perform better than the other two classifiers.

Table 4. Comparisons of malware detection effectiveness between IGIA and other classification methods

Training	TP	FP	TN	FN	ACY
IGIA	1,533	254	7,433	7	**0.9717**
SVM	926	158	7,529	614	0.9163
DT	1,021	708	6,979	519	0.8670
Testing	TP	FP	TN	FN	ACY
IGIA	65	55	1,872	26	**0.9598**
SVM	54	232	1,695	37	0.8667
DT	57	297	1,630	34	0.8360

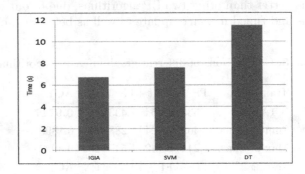

Fig. 3. Comparisons of malware detection efficiency between the proposed graph inference algorithm (IGIA) and other classification methods

6   Conclusion

In this paper, we provide deep analysis of malware's social relation network and study how it can be used for malware detection; we also propose an effective graph inference algorithm (IGIA) for malware detection based on the constructed file relation graphs. Empirical studies on large and real data collection obtained from Comodo Cloud Security Center illustrate that our proposed method outperforms other alternate data mining based approaches in malware detection. In our future work, we will further investigate more social relation based features (e.g., file co-operations) for malware detection and analyze the properties of different kinds of social relations among file samples. We will also design a full detection solution by combining social relations with other detection features to further improve the detection accuracy.

Acknowledgments. The authors would also like to thank the anti-malware experts of Comodo Security Lab for the data collection as well as helpful discussions and supports. The work of S. Hou, Lingwei Chen, and Y. Ye is supported by the U.S. National Science Foundation under grant CNS-1618629; the work of Lifei Chen is supported by the Chinese National Science Foundation under grant 61672157.

References

1. Bailey, M., Oberheide, J., Andersen, J., Mao, Z.M., Jahanian, F., Nazario, J.: Automated classification and analysis of internet malware. In: Kruegel, C., Lippmann, R., Clark, A. (eds.) RAID 2007. LNCS, vol. 4637, pp. 178–197. Springer, Heidelberg (2007). https://doi.org/10.1007/978-3-540-74320-0_10
2. Bayer, U., Comparetti, P.M., Hlauschek, C., Kruegel, C., Kirda, E.: Scalable, behavior-based malware clustering. In: NDSS 2009 (2009)
3. Bayer, U., Moser, A., Kruegel, C., Kirda, E.: Dynamic analysis of malicious code. J. Comput. Virol. **2**(1), 67–77 (2006)
4. Beaucamps, P., Filiol, E.: Malware pattern scanning schemes secure against black box analysis. J. Comput. Virol. **2**(1), 35–50 (2006)
5. Beaucamps, P., Filiol, E.: On the possibility of practically obfuscating programs towards a unified perspective of code protection. J. Comput. Virol. **3**(1), 3–21 (2007)
6. Bishop, C.: Pattern Recognition and Machine Learning. Information Science and Statistics. Springer, New York (2006)
7. Chau, D.H., Nachenberg, C., Wilhelm, J., Wright, A., Faloutsos, C.: Polonium: tera-scale graph mining for malware detection. In: SIAM International Conference on Data Mining (SDM) (2011)
8. Chen, L., Hardy, W., Ye, Y., Li, T.: Analyzing file-to-file relation network in malware detection. In: Wang, J., Cellary, W., Wang, D., Wang, H., Chen, S.-C., Li, T., Zhang, Y. (eds.) WISE 2015. LNCS, vol. 9418, pp. 415–430. Springer, Cham (2015). https://doi.org/10.1007/978-3-319-26190-4_28
9. Chen, L., Li, T., Abdulhayoglu, M., Ye, Y.: Intelligent malware detection based on file relation graphs. In: Proceedings of the 9th IEEE International Conference on Semantic Computing, pp. 85–92 (2015)
10. Clarkson, K.L.: Large-scale malware analysis, detection, and signature generation. Ph.D. dissertation, University of Michigan (2011)
11. CSI: 12th annual edition of the CSI computer crime and security survey. Technical report, Computer Security Institute (2007)
12. Filiol, E., Jacob, G., Liard, M.: Evaluation methodology and theoretical model for antiviral behavioural detection strategies. J. Comput. Virol. **3**(1), 27–37 (2007)

13. Jung, W., Kim, S., Choi, S.: Poster: deep learning for zero-day flash malware detection. In: S&P 2015 (2015)
14. 2013–2014 Internet security report in China (2014). http://www.ijinshan.com/news/2014011401.shtml
15. Kolter, J.Z., Maloof, M.A.: Learning to detect malicious executables in the wild. In: KDD 2004, pp. 470–478 (2004)
16. Li, Y., Ma, R., Jiao, R.: A hybrid malicious code detection method based on deep learning. IJSIA **9**, 205–216 (2015)
17. McGlohon, M., Bay, S., Anderle, M.G., Steier, D.M., Faloutsos, C.: Snare: a link analytic system for graph labeling and risk detection. In: KDD 2009 (2009)
18. Noorshams, N., Wainwright, M.J.: Belief propagation for continuous state spaces: stochastic message-passing with quantitative guarantees. J. Mach. Learn. Res. **14**(1), 2799–2835 (2013)
19. Pearl, J.: Reverend bayes on inference engines: a distributed hierarchical approach. In: Proceedings of the Second National Conference on Artificial Intelligence, pp. 133–136 (1982)
20. Reddy, D.K.S., Pujari, A.K.: N-gram analysis for computer virus detection. J. Comput. Virol. **2**(3), 231–239 (2006)
21. Shah, S., Jani, H., Shetty, S., Bhowmick, K.: Virus detection using artificial neural networks. Int. J. Comput. Appl. **84**, 17–23 (2013)
22. Tamersoy, A., Roundy, K., Chau, D.H.: Guilt by association: large scale malware detection by mining file-relation graphs. In: KDD 2014 (2014)
23. Zeus: a persistent criminal enterprise (2010). http://www.trendmicro.com/cloudcontent/us/pdfs/security-intelligence/white-papers/wpzeuspersistent-criminal-enterprise.pdf
24. Venzhega, A., Zhinalieva, P., Suboch, N.: Graph-based malware distributors detection. In: WWW 2013 (2013)
25. Wang, J., Deng, P., Fan, Y., Jaw, L., Liu, Y.: Virus detection using data mining techniques. In: ICDM (2003)
26. Ye, Y., Li, T., Chen, Y., Jiang, Q.: Automatic malware categorization using cluster ensemble. In: KDD 2010, pp. 95–104 (2010)
27. Ye, Y., Li, T., Zhu, S., Zhuang, W., Tas, E., Gupta, U., Abdulhayoglu, M.: Combining file content and file relations for cloud based malware detection. In: KDD 2011, pp. 222–230 (2011)
28. Ye, Y., Li, T., Jiang, Q., Han, Z., Wan, L.: Intelligent file scoring system for malware detection from the gray list. In: KDD 2009 (2009)
29. Ye, Y., Wang, D., Ye, D.: IMDS: intelligent malware detection system. In: KDD 2007, pp. 1043–1047 (2007)
30. Yedidia, J.S., Freeman, W.T., Weiss, Y.: Understanding belief propagation and its generalizations. Morgan Kaufmann Publishers Inc., San Francisco, CA, USA (2003)

An Adversarial Machine Learning Model Against Android Malware Evasion Attacks

Lingwei Chen[1], Shifu Hou[1], Yanfang Ye[1(✉)], and Lifei Chen[2]

[1] Department of Computer Science and Electrical Engineering,
West Virginia University, Morgantown, WV 26506, USA
{lgchen,shhou}@mix.wvu.edu, yanfang.ye@mail.wvu.edu
[2] School of Mathematics and Computer Science, Fujian Normal University,
Fuzhou 350117, Fujian, China
clfei@fjnu.edu.cn

Abstract. With explosive growth of Android malware and due to its damage to smart phone users, the detection of Android malware is one of the cybersecurity topics that are of great interests. To protect legitimate users from the evolving Android malware attacks, systems using machine learning techniques have been successfully deployed and offer unparalleled flexibility in automatic Android malware detection. Unfortunately, as machine learning based classifiers become more widely deployed, the incentive for defeating them increases. In this paper, we explore the security of machine learning in Android malware detection on the basis of a learning-based classifier with the input of Application Programming Interface (API) calls extracted from the smali files. In particular, we consider different levels of the attackers' capability and present a set of corresponding evasion attacks to thoroughly assess the security of the classifier. To effectively counter these evasion attacks, we then propose a robust secure-learning paradigm and show that it can improve system security against a wide class of evasion attacks. The proposed model can also be readily applied to other security tasks, such as anti-spam and fraud detection.

Keywords: Adversarial machine learning · Android malware detection · Evasion attack

1 Introduction

Due to their mobility and ever expanding capabilities, smart phones have been widely used to perform the tasks, such as banking and automated home control, in people's daily life. In recent years, there has been an exponential growth in the number of smart phone users around the world and it is estimated that 77.7% of all devices connected to the Internet will be smart phones in 2019 [13]. Designed as an open, free, and programmable operation system, Android as one of the most popular smart phone platforms dominates the current market share [1]. However, the openness of Android not only attracts the developers for

S. Song et al. (Eds.): APWeb-WAIM 2017 Workshops, LNCS 10612, pp. 43–55, 2017.
https://doi.org/10.1007/978-3-319-69781-9_5

producing legitimate applications (apps), but also attackers to deliver malware (short for *ma*licious soft*ware*) onto unsuspecting users to disrupt the mobile operations. Today, a lot of android malware (e.g., Geinimi, DriodKungfu and Hongtoutou) is released on the markets [25], which poses serious threats to smart phone users, such as stealing user information and sending SMS advertisement spams without the user's permission [9]. According to Symantec's recent Internet Security Threat Report [21], one in every five Android apps were actually malware. To protect legitimate users from the attacks of Android malware, intelligent systems using machine learning techniques have been successfully developed in recent years [11,12,23,24,26,29]. Though these machine learning techniques offer exceptional performance in automatic Android malware detection, machine learning itself may open the possibility for an adversary who maliciously "mistrains" a classifier (e.g., by changing data distribution or feature importance) in a detection system. When the learning system is deployed in a real-world environment, it is of a great interest for the attackers to actively manipulate the data to make the classifier producing minimum true positive (i.e., maximumly misclassifying Android malware as benign).

Android malware attackers and anti-malware defenders are actually engaged in a never-ending arms race, where both the attackers and defenders analyze the vulnerabilities of each other, and develop their own optimal strategies to overcome the opponents [5,18]. For example, attackers use repackaging and obfuscation to evade the anti-malware venders' detection. Though, the issues of machine learning security are starting to be leveraged [3,4,6,7,16,19,20,30], most existing researches for adversarial machine learning focus in the area of spam email detection, but rarely for Android malware detection. However, with the popularity of machine learning based detections, such adversaries will sooner or later present. In this paper, with the inputs of Application Programming Interface (API) calls extracted from the smali files (smali is an assemble/dissembler for the Dalvid executable (dex) files and provides readable code in smali language), we explore the security of machine learning in Android malware detection on the basis of a learning-based classifier. The major contributions of our work can be summarized as follows:

- *Thorough exploration of Android malware evasion attacks under different scenarios:* The attacker may have different levels of knowledge of the targeted learning system [19]. We define a set of evasion attacks corresponding to the different scenarios, and implement a thorough security analysis of a learning-based classifier.
- *An adversarial machine learning model against the evasion attacks:* Based on the learning system tainted by the attacks, we present an adversarial learning model against the evasion attacks in Android malware detection, in which we incorporate evasion data manipulation into the learning algorithm and enhance the robustness of the classifier using the security regularization term over the evasion cost.
- *Comprehensive experimental study on a real sample collection from an anti-malware industry company:* We collect the sample set from Comodo Cloud

Security Center, and provide a series of comprehensive experiments to empirically access the performances of our proposed methods.

The rest of the paper is organized as follows. Section 2 defines the problem of machine learning based Android malware detection. Section 3 describes the evasion attacks under different scenarios and their corresponding implementations. Section 4 introduces an adversarial learning model against the evasion attacks. Section 5 systematically evaluates the effectiveness of the proposed methods. Section 6 discusses the related work. Finally, Sect. 7 concludes.

2 Machine Learning Based Android Malware Detection

Based on the collected Android apps, without loss of generality, in this paper, we extract API calls from the smali files as the features, since they are used by the apps in order to access operating system functionality and system resources and thus can be used as representations of the behaviors of an app [12]. To extract the API calls, the Android app is first unzipped to provide the dex file, and then the dex file is further decompiled into smali code (i.e., the interpreted, intermediate code between Java and DalvikVM [8]) using a well-known reverse engineering tool APKTool [2]. The converted smali code can then be parsed for API call extraction. For example, the API calls of *"Lorg/apache/http/HttpRequest;* $\rightarrow containsHeader"$ and *"Lorg/apache/ http/ HttpRequest; $\rightarrow addHeader"$* can be extracted from "winterbird.apk" (MD5: *53cec6444101d19 76af1b253ff5b2226*) which is a theme wallpaper app embedded with malicious code that can steal user's credential. Note that other feature representations, either static or dynamic, are also applicable in our further investigation. Resting on the extracted API calls, we denote our dataset D to be of the form $D = \{\mathbf{x}_i, y_i\}_{i=1}^n$ of n apps, where $\mathbf{x}_i$ is the set of API calls extracted from app i, and y_i is the class label of app i, where $y_i \in \{+1, -1, 0\}$ (+1 denotes malicious, -1 denotes benign, and 0 denotes unknown). Let d be the number of all extracted API calls in the dataset D. Each of the app can be represented by a binary feature vector:

$$\mathbf{x}_i = <x_{i1}, x_{i2}, x_{i3}, ..., x_{id}>, \tag{1}$$

where $\mathbf{x}_i \in \mathbb{R}^d$, and $x_{ij} = \{0, 1\}$ (i.e., if app i includes API_j, then $x_{ij} = 1$; otherwise, $x_{ij} = 0$).

The problem of machine learning based Android malware detection can be stated in the form of: $f : \mathcal{X} \rightarrow \mathcal{Y}$ which assigns a label $y \in \mathcal{Y}$ (i.e., -1 or $+1$) to an input app $\mathbf{x} \in \mathcal{X}$ through the learning function f. A general linear classification model for Android malware detection can be thereby denoted as:

$$\mathbf{f} = \text{sign}(f(\mathbf{X})) = \text{sign}(\mathbf{X}^T \mathbf{w} + \mathbf{h}), \tag{2}$$

where $\mathbf{f}$ is a vector, each of whose elements is the label (i.e., malicious or benign) of an app to be predicted, each column of matrix $\mathbf{X}$ is the API feature vector of an app, $\mathbf{w}$ is the coefficients and $\mathbf{h}$ is the biases. More specifically, the machine learning on the basis of a linear classifier can be formalized as an optimization problem [28]:

$$\underset{\mathbf{f},\mathbf{w},\mathbf{h};\boldsymbol{\xi}}{\operatorname{argmin}} \frac{1}{2}||\mathbf{y} - \mathbf{f}||^2 + \frac{1}{2\beta}\mathbf{w}^T\mathbf{w} + \frac{1}{2\gamma}\mathbf{h}^T\mathbf{h} + \boldsymbol{\xi}^T(\mathbf{f} - \mathbf{X}^T\mathbf{w} - \mathbf{h}) \qquad (3)$$

subject to Eq. 2, where $\mathbf{y}$ is the labeled information vector, $\boldsymbol{\xi}$ is Lagrange multiplier, β and γ are the regularization parameters. Note that Eq. 3 is a typical linear classification model consisting of specific loss function and regularization terms. Without loss of generality, the equation can be transformed into different linear models depending on the choices of loss function and regularization terms.

3   Implementation of Evasion Attacks

Considering that the attackers may know differently about: (i) the feature space, (ii) the training sample set, and (iii) the learning algorithm [19], we characterize their knowledge in terms of a space Γ that encodes knowledge of the feature space X, the training sample set D, and the classification function f. To this end, we present three well-defined evasion attacks to facilitate our analysis: (1) *Mimicry Attacks*: The attackers are assumed to know the feature space and be able to obtain a collection of apps to imitate the original training dataset, i.e., $\Gamma = (X, \hat{D})$. (2) *Imperfect-knowledge Attacks*: In this case, it's assumed that both the feature space and the original training sample set can be fully controlled by the attackers, i.e., $\Gamma = (X, D)$. (3) *Ideal-knowledge Attacks*: This is the worst case where the learning algorithm is also known to the attackers, i.e., $\Gamma = (X, D, f)$, allowing them to perfectly access to the learning system.

3.1   Evasion Cost

Evasion attacks are universally modeled as an optimization problem: given an original malicious app $\mathbf{x} \in \mathcal{X}^+$, the evasion attacks attempt to manipulate it to be detected as benign (i.e., $\mathbf{x}' \in \mathcal{X}^-$), with the minimal evasion cost. Considering API calls of each app is formatted as a binary vector, the cost of feature manipulation (addition or elimination), can be encoded as the distance between $\mathbf{x}$ and $\mathbf{x}'$:

$$c(\mathbf{x}', \mathbf{x}) = ||\mathbf{a}^T(\mathbf{x}' - \mathbf{x})||_p^p, \qquad (4)$$

where p is a real number and $\mathbf{a}$ is a weight vector, each of which denotes the relative cost of changing a feature. The cost function can be considered as ℓ_1-norm or ℓ_2-norm depending on the feature space. For the attackers, to evade the detection, adding or hiding some API calls in a malicious app does not seem difficult. However, some specific API calls may affect the structure for intrusive functionality, which may be more expensive to be modified. Therefore we view the evasion cost c practically significant. Accordingly, there is an upper limit of the maximum manipulations that can be made to the original malicious app $\mathbf{x}$. Therefore, the manipulation function $\mathcal{A}(\mathbf{x})$ can be formulated as

$$\mathcal{A}(\mathbf{x}) = \begin{cases} \mathbf{x}' & \text{sign}(f(\mathbf{x}')) = -1 \ and \ c(\mathbf{x}', \mathbf{x}) \le \delta_{\max} \\ \mathbf{x} & \text{otherwise} \end{cases}, \tag{5}$$

where the malicious app is manipulated to be misclassified as benign only if the evasion cost is less than or equal to a maximum cost $\delta_{\max}$. Let $\mathbf{f}' = \text{sign}(f(\mathcal{A}(\mathbf{X})))$, then the main idea for an evasion attack is to maximize the total loss of classification (i.e., $\text{argmax} \ \frac{1}{2}\|\mathbf{y} - \mathbf{f}'\|^2$), which means that the more malicious Android apps are misclassified as benign, the more effective the evasion attack could be. An ideal evasion attack modifies a small but optimal portion of features of the malware with minimal evasion cost, while makes the classifier achieve lowest true positive rate.

3.2   Evasion Attack Method

To implement the evasion attack, it's necessary for the attackers to choose a relevant subset of API calls applied for feature manipulations. To evade the detection with lower evasion cost, the attackers may inject the API calls most relevant to benign apps while remove the ones more relevant to malware. To stimulate the attacks, we rank each API call using Max-Relevance algorithm [17] which has been successfully applied in malware detection [27]. Based on a real sample collection of $2,334$ Android apps ($1,216$ malicious and $1,118$ benign) obtained from Comodo Cloud Security Center, $1,022$ API calls are extracted. Figure 1 shows the Cumulative Distribution Function (CDF) of the API calls' relevance scores, from which we can see that (1) for different API calls, some are explicitly relevant to malware, while some have high influence on benign apps; and (2) those API calls with extremely low relevance scores (below 0.06 in our case) have limited or no contributions in malware detection, thus we exclude them for feature manipulations. Therefore, we rank each API call and group them into two sets for feature manipulations: $\mathcal{M}$ (those highly relevant to malware) and $\mathcal{B}$ (those highly relevant to benign apps) in the descent order of $I(x, +1)$ and $I(x, -1)$ respectively.

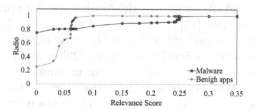

Fig. 1. Relevance score CDF of API calls

To evaluate the capability of an evasion attack, we further define a function $g(\mathcal{A}(\mathbf{X}))$ to represent the accessibility to the learning system by the attack:

$$g(\mathcal{A}(\mathbf{X})) = ||\mathbf{y} - \mathbf{f}'||^2, \tag{6}$$

which implies the number of malicious apps being misclassified as benign. The underlying idea of our evasion attack *EvAttack* is to perform feature manipulations with minimum evasion cost while maximize the total loss of classification in Eq. (6). Specifically, we conduct bi-directional feature selection, that is, forward feature addition performed on $\mathcal{B}$ and backward feature elimination performed on $\mathcal{M}$. At each iteration, an API call will be selected for addition or elimination depending on the fact how it influences the value of $g(\mathcal{A}(\mathbf{X}))$. The evasion attack $\boldsymbol{\theta} = \{\boldsymbol{\theta}^+, \boldsymbol{\theta}^-\}$ will be drawn from the iterations, where $\boldsymbol{\theta}^+, \boldsymbol{\theta}^- \in \{0,1\}^d$ (if API_i is selected for elimination (or addition), then $\theta_i^+(\theta_i^-) = 1$; otherwise, $\theta_i^+(\theta_i^-) = 0$). The iterations will end at the point where the evasion cost reaches to maximum ($\delta_{\max}$), or the features available for addition and elimination are all manipulated. Given $m = \max(|\mathcal{M}|, |\mathcal{B}|)$, *EvAttack* requires $O(n_t m(\mu^+ + \mu^-))$ queries, in which n_t is the number of malicious apps that the attackers want to evade the detection, μ^+ and μ^- are the numbers of selected features for elimination and addition respectively ($\mu^+ \ll d$, $\mu^- \ll d$, $m \ll d$).

4   Adversarial Machine Learning Model Against Evasion Attacks

To be resilient against the evasion attacks, an anti-malware defender may modify features of the training dataset, or analyze the attacks and retrain the classifier accordingly to counter the adversary's strategy [18]. However, these methods typically suffer from assumption of specific attack models or substantial modifications of learning algorithms [15]. In this paper, we provide a systematic model to formalize the impact of the evasion attacks with respect to system security and robustness in Android malware detection. To this end, we perform our security analysis of the learning-based classifier resting on the application setting that the defender draws the well-crafted *EvAttack* from the observed sample space, since the attack is modeled as optimization under generic framework in which the attackers try to (1) maximize the number of malicious apps being classified as benign, and (2) minimizes the evasion cost for optimal attacks over the learning-based classifier [15].

Incorporating the evasion attack $\boldsymbol{\theta}$ into the learning algorithm can be performed through computing a Stackelberg game [20] or adding adversarial samples [15], which enables us to provide a significant connection between training and the adversarial action. An adversarial learning model for Android malware detection is supposed to be more resilient to the evasion attacks. It's recalled that an optimal evasion attack aims to manipulate a subset of the features with minimum evasion cost while maximize the total loss of classification. In contrast, to secure the classifier in Android malware detection, we would maximize the evasion cost for the attacks [30]: from the analysis of the adversary problem [4,14], we can find that the larger the evasion cost, the more manipulations need to be performed, and the more difficult the attack is. Therefore, in our proposed adversarial learning model, we will not only incorporate the evasion attack θ into the learning algorithm, but also enhance the classifier by using a security regularization term based on the evasion cost.

We first define the resilience coefficient of a classifier as:

$$s(\mathcal{A}(\mathbf{x}), \mathbf{x}) = 1/c(\mathcal{A}(\mathbf{x}), \mathbf{x}), \tag{7}$$

subject to Eq. 4, which is converse to the evasion cost. We then define a diagonal matrix for the adversary action denoted as $\mathbf{S} \in \mathbb{R}^{n \times n}$, where the diagonal element $S_{ii} = s(\mathcal{A}(\mathbf{x}_i), \mathbf{x}_i)$ and the remaining elements in the matrix are 0. Based on the concept of label smoothness, we can secure the classifier with the constraint as $\mathbf{f}'^T \mathbf{S} \mathbf{y}$. Since the learning based Android malware detection can be formalized as an optimization problem denoted by Eq. 3, we can then bring a security regularization term to enhance its security. This constraint penalizes parameter choices, smooths the effects the attack may cause, and in turn helps to promote the optimal solution for the local minima in the optimization problem. Therefore, to minimize classifier sensitivity to feature manipulation, we can minimize the security regularization term. Based on Eq. 3, we can formulate an adversarial learning model against evasion attacks as:

$$\underset{\mathbf{f}', \mathbf{w}, \mathbf{h}; \boldsymbol{\xi}}{\operatorname{argmin}} \frac{1}{2}\|\mathbf{y} - \mathbf{f}'\|^2 + \frac{\alpha}{2}\mathbf{f}'^T \mathbf{S} \mathbf{y} + \frac{1}{2\beta}\mathbf{w}^T \mathbf{w} + \frac{1}{2\gamma}\mathbf{h}^T \mathbf{h} + \boldsymbol{\xi}^T(\mathbf{f}' - \mathbf{X}'^T \mathbf{w} - \mathbf{h}) \tag{8}$$

subject to $\mathbf{f}' = \operatorname{sign}(f(\mathcal{A}(\mathbf{X})))$ and $\mathbf{X}' = \mathcal{A}(\mathbf{X})$, where α is the regularization parameter for the security constraint. To solve the problem in Eq. 8, let

$$\mathcal{L}(\mathbf{f}', \mathbf{w}, \mathbf{h}; \boldsymbol{\xi}) = \frac{1}{2}\|\mathbf{y} - \mathbf{f}'\|^2 + \frac{\alpha}{2}\mathbf{f}'^T \mathbf{S} \mathbf{y} + \frac{1}{2\beta}\mathbf{w}^T \mathbf{w} + \frac{1}{2\gamma}\mathbf{h}^T \mathbf{h} + \boldsymbol{\xi}^T(\mathbf{f}' - \mathbf{X}'^T \mathbf{w} - \mathbf{h}).$$
$$\tag{9}$$

As $\frac{\partial \mathcal{L}}{\partial \mathbf{w}} = 0$, $\frac{\partial \mathcal{L}}{\partial \mathbf{h}} = 0$, $\frac{\partial \mathcal{L}}{\partial \boldsymbol{\xi}} = 0$, $\frac{\partial \mathcal{L}}{\partial \mathbf{f}'} = 0$, we have

$$\mathbf{w} = \beta \mathbf{X}' \boldsymbol{\xi}, \tag{10}$$

$$\mathbf{h} = \gamma \boldsymbol{\xi}, \tag{11}$$

$$\mathbf{f}' = \mathbf{X}'^T \mathbf{w} + \mathbf{h}, \tag{12}$$

$$\mathbf{f}' = \mathbf{y} - \frac{\alpha}{2}\mathbf{S} \mathbf{y} - \boldsymbol{\xi}. \tag{13}$$

Based on the derivation from Eqs. 10, 11 and 12, we have

$$\boldsymbol{\xi} = (\beta \mathbf{X}'^T \mathbf{X}' + \gamma \mathbf{I})^{-1} \mathbf{f}'. \tag{14}$$

We substitute Eqs. 14 to 13, then we can get our adversarial model in malware detection (*AdvMD*) as:

$$((\beta \mathbf{X}'^T \mathbf{X}' + \gamma \mathbf{I}) + \mathbf{I})\mathbf{f}' = (\mathbf{I} - \frac{\alpha}{2}\mathbf{S})(\beta \mathbf{X}'^T \mathbf{X}' + \gamma \mathbf{I})\mathbf{y}. \tag{15}$$

Since the size of $\mathbf{X}'$ is $d \times n$, the computational complexity for Eq. 15 is $O(n^3)$. If $d < n$, we can follow Woodbury identity [22] to transform Eqs. 15 to 16 and reduce the complexity to $O(d^3)$.

$$(\mathbf{I} + \gamma^{-1}\mathbf{I} - \gamma^{-1}\mathbf{X}'^T(\gamma\beta^{-1}\mathbf{I} + \mathbf{X}'\mathbf{X}'^T)^{-1}\mathbf{X}')\mathbf{f}' = (\mathbf{I} - \frac{\alpha}{2}\mathbf{S})\mathbf{y}. \tag{16}$$

To solve the above secure learning in Eq. 15, conjugate gradient descent method can be applied, which is also applicable to Eq. 16, provided that all the variables are configured as the correct initializations.

5   Experimental Results and Analysis

In this section, three sets of experiments are conducted to evaluate our proposed methods. The real sample collection obtained from Comodo Cloud Security Center contains $2,334$ apps ($1,216$ malicious and $1,118$ benign) with $1,022$ extracted API calls. In our experiments, we use k-fold cross-validations for performance evaluations. We evaluate the detection performance of different methods using *TP* (true positive), *TN* (true negative), *FP* (false positive), *FN* (false negative), *TPR* (TP rate), *FPR* (FP rate), *FNR* (FN rate), and *ACC* (accuracy).

According to the relevance scores analyzed in Sect. 3.2, those API calls whose relevance scores are lower than the empirical threshold (i.e., 0.06 in our application) will be excluded from consideration. Therefore, $|\mathcal{M}| = 199$, $|\mathcal{B}| = 333$. Considering different weights for API calls, we optimize the weight for each feature ($a \in [0, 1]$). We exploit the average number of API calls that each app possesses (i.e., 127) to define δ_{max} based on the CDF drawn from all these numbers. When we set δ_{max} as 20% of the average number of API calls that each app possesses (i.e., 25), the average feature manipulation of each app is about 10% of its extracted API calls, which could be considered as a reasonable trade-off to conduct the evasion attacks considering the feature number and manipulation cost. Therefore, we run our evaluation of the proposed evasion attacks with $\delta_{max} = 25$.

5.1   Evaluation of *EvAttack* Under Different Scenarios

Given $\delta_{max} = 25$, we first evaluate *EvAttack* under different scenarios: (1) In mimicry (MMC) attack, i.e., $\Gamma = (X, \hat{D})$, we select 200 random apps (100 malware and 100 benign apps) from our collected sample set (excluding those ones for testing) as our mimic dataset and utilize linear Support Vector Machine (SVM) as the surrogate classifier to train these app samples. (2) In imperfect-knowledge (IPK) attack, i.e., $\Gamma = (X, D)$, we implement this attack in a consistent manner as MMC attack where the only difference is that we apply 90%

of the collected apps to train SVM. (3) In Ideal-knowledge (IDK) attack, i.e., $\Gamma = (X, D, f)$, we conduct *EvAttack* based on all the collected apps and the learning model in Eq. 3. Note that, *EvAttack* is applied to all these scenarios resting on the same $\delta_{\max}$, and each attack is performed by 10-fold cross-validation. The experimental results are shown in Table 1, which illustrate that the performance of the attack significantly relies on the available knowledge the attackers have. In ideal-knowledge scenarios, the *FNR* of IDK reaches to 0.7227 (i.e., 72.27% of testing malicious apps are misclassified as benign) that is superior to MMC and IPK.

Table 1. Evaluation of the evasion attacks under different scenarios

Scenarios	TP	FN	ACC	FNR
Before attack	111	8	96.12%	0.0672
MMC attack	85	34	84.91%	0.2857
IPK attack	48	71	68.97%	0.5966
IDK attack	33	86	62.50%	0.7227

5.2 Comparisons of *EvAttack* and Other Attacks

We further compare *EvAttack* with other attack methods including: (1) only injecting API calls from $\mathcal{B}$ (*Method 1*); (2) only eliminating API calls from $\mathcal{M}$ (*Method 2*); (3) injecting $(1/2 \times \delta_{\max})$ API calls from $\mathcal{B}$ and eliminating $(1/2 \times \delta_{\max})$ API calls from $\mathcal{M}$ (*Method 3*); (4) simulating anonymous attack by randomly manipulating API calls for addition and elimination (*Method 4*). The experimental results which average over the 10-fold cross-validations shown in Fig. 2 demonstrate that the performances of the attacks vary when using different feature manipulation methods (*Method 0* is the baseline before attack and *Method 5* denotes the *EvAttack*) with the same evasion cost $\delta_{\max}$: (1) *Method 2* performs worst with the lowest *FNR* which denotes that elimination is not as effective as others; (2) *Method 3* performs better than the methods only applying feature addition or elimination, and the anonymous attack, due to its bi-directional feature manipulation over $\mathcal{B}$ and $\mathcal{M}$; (3) *EvAttack* can greatly improve the *FNR* to 0.7227 and degrade the detection accuracy *ACC* to 62.50%, which outperforms other four feature manipulation methods for its well-crafted attack strategy.

5.3 Evaluation of *AdvMD* Against Evasion Attacks

In this set of experiments, we validate the effectiveness of *AdvMD* to the well-crafted attacks. We use *EvAttack* to taint the malicious apps in the testing set, and access the performances in different ways: (1) the baseline before attack;

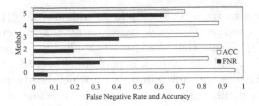

Fig. 2. *FNR*s and accuracies of different attacks.

(2) the classifier under attack; (3) the classifier retrained using the updated training dataset [15,20]; (4) *AdvMD*. We conduct the 10-fold cross-validations, experimental results of different learning models are shown in Fig. 3. Figure 3(a) shows the comparisons of FPR, TPR and ACC for different learning models before/against *EvAttack*, and Fig. 3(b) illustrates the ROC curves of different learning models before/against *EvAttack*. From Fig. 3, we can observe that (i) the retraining techniques can somehow defense the evasion attacks, but the performance still remain unsatisfied; while (ii) *AdvMD* can significantly improve the *TPR* and *ACC*, and bring the malware detection system back up to the desired performance level, the accuracy of which is 91.81%, approaching the detection results before the attack. We also implement the anonymous attack by randomly selecting the features for manipulation. Under the anonymous attack, *AdvMD* has zero knowledge of what the attack is. Even in such case, *AdvMD* can still improve the *TPR* from 0.7815 to 0.8367. Based on these properties, *AdvMD* can be a resilient solution in Andriod malware detection.

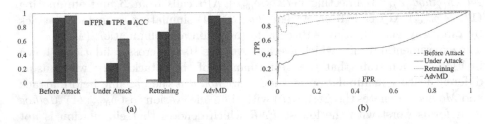

Fig. 3. Comparisons of different learning models before/against *EvAttack*.

6   Related Work

Adversarial machine learning problems are starting to be leveraged from either adversarial or defensive perspectives in some domains, such as anti-spam and intrusion detection. Lowd and Meek [16] introduced an ACRE framework to study how an adversary can learn sufficient information from the features to construct targeted attacks using minimal adversarial cost. Zhang et al. [30],

Li et al. [15], and Biggio et al. [4] took gradient steps to find the closest evasion point $\mathbf{x}'$ to the malicious sample $\mathbf{x}$. Haghtalab et al. [10] proposed to learn the behavioral model of a bounded rational attacker by observing how the attacker responded to three defender strategies. To combat the evasion attacks, ample research efforts have been devoted to the security of machine learning. Wang et al. [20] modeled the adversary action as it controlling a vector $\boldsymbol{\alpha}$ to modify the training dataset $\mathbf{x}$. Debarr et al. [7] explored randomization to generalize learning model to estimate some parameters that fit the data best. Kolcz and Teo [14] investigated a feature reweighting technique to avoid single feature over-weighting. More recently, robust feature selection methods have also been proposed to counter some kinds of evasion data manipulations [20,30]. However, most of these works rarely investigate the security of machine learning in Android malware detection. Different from the existing works, we explore the adversarial machine learning in Android malware detection by providing a set of evasion attacks to access the security of the classifier over different capabilities of the attackers and enhancing the learning algorithm using evasion action and security regularization term.

7   Conclusion and Future Work

In this paper, we take insights into the machine learning based model and its evasion attacks. Considering different knowledge of the attackers, we implement an evasion attack *EvAttack* under three scenarios by manipulating an optimal portion of the features to evade the detection. Accordingly, an adversarial learning model *AdvMD*, enhanced by evasion data manipulation and security regularization term, is presented against these attacks. Three sets of experiments based on the real sample collection from Comodo Cloud Security Center are conducted to empirically validate the proposed approaches. The experimental results demonstrate that *EvAttack* can greatly evade the detection, while *AdvMD* can be a robust and practical solution against the evasion attacks in Android malware detection. In our future work, we will further explore the poisoning attacks in which the attackers alter the training precess through influence over the training data, as well as its resilient detection.

Acknowledgments. The authors would also like to thank the experts of Comodo Security Lab for the data collection and helpful discussions. The work is partially supported by the U.S. National Science Foundation under grant CNS-1618629 and Chinese NSF grant 61672157.

References

1. Android: iOS combine for 91 percent of market. http://www.cnet.com
2. APKTool. http://ibotpeaches.github.io/Apktool/
3. Barreno, M., Nelson, B., Sears, R., Joseph, A.D., Tygar, J.D.: Can machine learning be secure? In: ASIACCS (2006)

4. Biggio, B., Fumera, G., Roli, F.: Evade hard multiple classifier systems. In: Okun, O., Valentini, G. (eds.) Applications of Supervised and Unsupervised Ensemble Methods. Studies in Computational Intelligence, pp. 15–38. Springer, Heidelberg (2009). doi:10.1007/978-3-642-03999-7_2
5. Biggio, B., Fumera, G., Roli, F.: Security evaluation of pattern classifiers under attack. IEEE TKDE **26**(4), 984–996 (2014)
6. Bruckner, M., Kanzow, C., Scheffer, T.: Static prediction games for adversarial learning problems. JMLR **13**, 2617–2654 (2012)
7. Debarr, D., Sun, H., Wechsler, H.: Adversarial spam detection using the randomized hough transform-support vector machine. In: ICMLA 2013, pp. 299–304 (2013)
8. Dex. http://www.openthefile.net/extension/dex
9. Felt, A.P., Finifter, M., Chin, E., Hanna, S., Wagner, D.: A survey of mobile malware in the wild. In: SPSM (2011)
10. Haghtalab, N., Fang, F., Nguyen, T.H., Sinha, A., Procaccia, A.D., Tambe, M.: Three strategies to success: learning adversary models in security games. In: IJCAI (2016)
11. Hou, S., Saas, A., Chen, L., Ye, Y.: Deep4MalDroid: a deep learning framework for android malware detection based on linux kernel system call graphs. In: WIW (2016)
12. Hou, S., Saas, A., Ye, Y., Chen, L.: DroidDelver: an android malware detection system using deep belief network based on API call blocks. In: Song, S., Tong, Y. (eds.) WAIM 2016. LNCS, vol. 9998, pp. 54–66. Springer, Cham (2016). doi:10.1007/978-3-319-47121-1_5
13. IDC. http://www.idc.com/getdoc.jsp?containerId=prUS25500515
14. Kolcz, A., Teo, C.H.: Feature weighting for improved classifier robustness. In: CEAS 2009 (2009)
15. Li, B., Vorobeychik, Y., Chen, X.: A general retraining framework for adversarial classification. In: NIPS 2016 (2016)
16. Lowd, D., Meek, C.: Adversarial learning. In: KDD, pp. 641–647 (2005)
17. Peng, H., Long, F., Ding, C.: Feature selection based on mutual information: criteria of max-dependency, max-relevance, and min-redundancy. IEEE Trans. Pattern Anal. Mach. Intell. **27**(8), 1226–1238 (2005)
18. Roli, F., Biggio, B., Fumera, G.: Pattern recognition systems under attack. In: Ruiz-Shulcloper, J., Sanniti di Baja, G. (eds.) CIARP 2013. LNCS, vol. 8258, pp. 1–8. Springer, Heidelberg (2013). doi:10.1007/978-3-642-41822-8_1
19. Šrndic, N., Laskov, P.: Practical evasion of a learning-based classifier: a case study. In: SP (2014)
20. Wang, F., Liu, W., Chawla, S.: On sparse feature attacks in adversarial learning. In: ICDM 2014 (2014)
21. Wood, P.: Internet Security Threat Report 2015. Symantec, California (2015)
22. Woodbury, M.A.: Inverting modified matrices. Statistical Research Group, Princeton University, Princeton, NJ (1950)
23. Wu, D., Mao, C., Wei, T., Lee, H., Wu, K.: DroidMat: android malware detection through manifest and API calls tracing. In: Asia JCIS (2012)
24. Wu, W., Hung, S.: DroidDolphin: a dynamic Android malware detection framework using big data and machine learning. In: RACS (2014)
25. Xu, J., Yu, Y., Chen, Z., Cao, B., Dong, W., Guo, Y., Cao, J.: MobSafe: cloud computing based forensic analysis for massive mobile applications using data mining. Tsinghua Sci. Technol. **18**, 418–427 (2013)

26. Yang, C., Xu, Z., Gu, G., Yegneswaran, V., Porras, P.: DroidMiner: automated mining and characterization of fine-grained malicious behaviors in android applications. In: Kutyłowski, M., Vaidya, J. (eds.) ESORICS 2014. LNCS, vol. 8712, pp. 163–182. Springer, Cham (2014). doi:10.1007/978-3-319-11203-9_10
27. Ye, Y., Li, D., Li, T., Ye, D.: IMDS: intelligent malware detection system. In: KDD 2007 (2007)
28. Ye, Y., Li, T., Zhu, S., Zhuang, W., Tas, E., Gupta, U., Abdulhayoglu, M.: Combining file content and file relations for cloud based malware detection. In: KDD 2011, pp. 222–230 (2011)
29. Yuan, Z., Lu, Y., Wang, Z., Xue, Y.: Droid-Sec: deep learning in android malware detection. In: SIGCOMM (2014)
30. Zhang, F., Chan, P.P.K., Biggio, B., Yeung, D.S., Roli, F.: Adversarial feature selection against evasion attacks. IEEE Trans. Cybern. 46(3), 766–777 (2015)

Word Similarity Computation with Extreme-Similar Method

Peiwen Du, Siding Chen, Xiaofei Xu, and Li Li[✉]

School of Computer and Information Science,
Southwest University, Chongqing 400715, China
irisdu1996@gmail.com, lily@swu.edu.cn

Abstract. Chinese word similarity calculation is a key technique in Chinese information processing. The most widely used word-based similarity calculations often fail to detect subtle differences between two words. This can lead to grossing mis-estimation of the similarity between two words. In this paper, we propose a new method to calculate the similarity between two Chinese words with a particular focus on comparing pairs of words which are very similar in meaning. A hybrid combination strategy is formulated incorporating other similarity calculations for scenarios between these two extreme conditions. Different corpora and models are used to train the proposed method, then combining with the score obtained from the Hownet and the final similarity value is refined accordingly. This model makes an important improvement to the existing strategies. Experiments on very similar words were conducted with two evaluation metrics, the Spearman and Pearson rank correlation coefficients. Our final results are 0.427/0.421 which outperforms the existing state-of-the-art models. It clearly shows the effectiveness of the proposed method.

Keywords: Chinese word similarity · Combination strategy · Hownet · Word2vector · Extreme-similar algorithm

1 Introduction

Chinese word similarity computation is a fundamental research in many Nature Language Processing (NLP) tasks. It provides a generic evaluation framework and is widely used in many fields, such as information retrieval, information extraction, text categorization, semantic divergence, case-based machine translation, and so on. Currently there are two methods to compute the similarity: knowledge-based and corpus-based methods [1].

Typically, knowledge-based word similarity [2] computation mainly relies on numerously manual semantic resources, such as semantic content or relative position within hierarchies. Obviously, it is a labor-intensive task that requires maintenance when new words and new word senses are coming out. And these methods have a disadvantage that is computing will work perfectly when the pair members are both in the lexicons. And the corpus-based methods overcome

© Springer International Publishing AG 2017
S. Song et al. (Eds.): APWeb-WAIM 2017 Workshops, LNCS 10612, pp. 56–65, 2017.
https://doi.org/10.1007/978-3-319-69781-9_6

these limitations by predicting unknown words. However, when facing large-scale corpora, it often leads to curse of dimensionality. Therefore, drawbacks have attracted increasing attention to integrate lexicons into word embedding to capture multiple semantics [3,4].

However, most of the existing methods suffer form the problem, which they can hardly compute similarity from extreme similar words. For instance, the words in the pair (购入gouru/buy, 购买goumai/buy, score = 7.6) got the score much lower than expectation and in the pair (消极xiaoji/negative, 积极jiji/positive, score = 8.6) got the higher one. In this paper, we propose a meaningful extreme-similar algorithm for Chinese word similarity computation.

The remainder of this article is organized as follows. Semantic similarity computation and distributed similarity computation are detailed in Sect. 2. Additionally in Sect. 2, we propose the extreme-similar algorithm to combine these two methods. In Sect. 3, we provide the experimental results and analysis. Finally, we discuss related topics and conclude in Sect. 4.

2 Methodology

2.1 Semantic Similarity Computation

As a knowledge system, hownet reflects the common and characteristics of the concept [5,6]. The size of the hownet depends largely on the size of the bilingual knowledge dictionary data file and it changes dynamically. Liu and Li [7] use set and structural characteristics to rewrite the word definition in hownet through structural method. They reviewed the similarity computation method in sememe (Broadly speaking, a sememe refers to the smallest basic semantic unit that cannot be reduced further),[1] set and structural characteristics and proposed a word similarity computation algorithm based on hownet. We leverage a method based on their work as shown in Eq. 1.

$$h_s(s_1, s_2) = \max_{i=1...n, j=1...m} h_s(s_{1i}, s_{2j}) \tag{1}$$

where $s_{1i}(i = 1...n)$ represents the sememe of the word1(s_1) and $s_{2j}(j = 1...n)$ represents the sememe of the word(s_2), furtermore $h_s(s_1, s_2)$ is the word similarity of s_1(word1) and s_2(word2) computed by the hownet-based algorithm. In order to facilitate the observation and integration of the experimental results, we transform original domain [0, 1] into [1, 10] via a simple function g(x) in Eq. 2:

$$g(x) = \begin{cases} 1, & x \leqslant 0 \\ 9x + 1, & x > 0 \end{cases} \tag{2}$$

---

[1] http://www.keenage.com/zhiwang/e_zhiwang_r.html.

2.2   Distributed Similarity Computation

Since most of the deep learning algorithm have large calculation cost, Mikolov *et al.* develops word2vec-based method which is a deep learning toolkit. Given an unlabeled training corpus, word2vec-based method produces a vector for each word in the corpus encoding with its semantic information [8]. There are two methods can be used to acquire the word vectors. The first is the singular value decomposition-based method, another is the iteration-based method. The latter one is utilized in our experiment. This method is mostly based on the training of language models to obtain the word vectors. As a result we are willing to use all models included to obtain the best vectors. The Continuous Bag-of-Word model (CBOW) and the skip-gram model (skip-gram) are shown in Fig. 1. After training, obtained vectors compose the vector spaces. We can define the semantic similarity of two words through using vector spaces by considering the cosine distance between them [9,10].

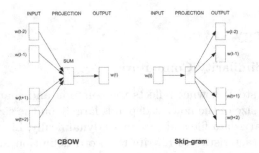

Fig. 1. The structure of the CBOW model

Continuous Bag of Words Model. The continuous bag of words model uses the given context to predict the probability of the target word. This calls for a modification to the neural network architecture. The modification is shown on the left of Fig. 1. Four words in the input layer is projected onto a projection layer, every dimension value of the projection is combined and weighted averaged, then a mean vector is constituted.

Skip-gram Model. The skip-gram Model reverses the use of target and context words. In this case, the target word is fed at the input, the hidden layer remains the same, and the output layer of the neural network is replicated multiple times to accommodate the chosen number of context words [11]. And it is shown on the right of Fig. 1.

In this paper we use the word2vec-based method to get the word vectors and we can define the similarity between two words by vectors cosine distance in Eq. 3:

$$simw\,(w_1, w_2) = \cos\theta = \frac{\boldsymbol{w}_1 \cdot \boldsymbol{w}_2}{\|\boldsymbol{w}_1\|\,\|\boldsymbol{w}_2\|} \tag{3}$$

Similarly, we transform original domain $[0, 1]$ into $[1, 10]$ by Eq. 2.

2.3   Our Extreme-Similar Algorithm

Due to traditional approaches pay little attention to the words that are extremely similar, some word pairs should have had a higher or lower score. We propose a simple but meaningful algorithm called extreme-similar algorithm. Obviously our approach focuses more on the pairs which are very similar.

Since labels in hownet-based method are noted manually, we have more confidence in the result of hownet-based model, which are defined as Simh. Also, Simw stands for the computing result of word2vec-based model.

If Simh is really high, it is more likely to say that this word pair is extremely similar. And when the difference between Simh and Simw is large, the average of two score is quiet likely lower than the reality, so we must refine the final score. In these situations, we set α and β as the algorithm threshold. Then we utilize sequence-similarity (Vsim1), which stands for the similarity between 2 Chinese words in the inspect of Chinese character sequence, and pattern-similarity (Vsim2), which measures the similar Chinese characters in a pair based on Tongyici Cilin, to refine the average (sim) of simh and simw. The Cilin dictionary is organized by a 5-layer hierarchical structure. Correspondingly, it supplies 5-layer patterns to encode for a group of words, which are joined by relationships like "synonym" or "relevance" [12]. Since the Geometric Mean is more closer to the bigger one, we calculate the square root (Vsim) of vsim1 and vsim2 as a reference to correct the Sim. In this way, Ssp is defined to characterize the adjustable space in a pair similarity measurement. The number 10 denotes two words in the pair are exactly the same, and moreover, 10 is the upper bound of Extreme-Similar.

The detailed steps are shown in Algorithm 1.

3   Data and Experimental Results

We present out experimental results in this section by first introducing the data and evaluation method, followed by the results of our system.

3.1   Dataset

We downloaded the Test Data from NLPCC-ICCPOL2016 Shared Task [13]. There are together 10,000 pairs of words in the file, in which only 500 pairs with labels that selected as the final test data. For the close of the online submission, we asked the organizers of NLPCC-ICCPOL for the golden human labelled data, which helps us to evaluate our experiments results. These words mainly from news articles and weibo text, which insure capturing word usages both in formal

---

Algorithm 1. Extreme-Similar Strategy

#both inputs of the function Hownet() and Word2vec() are the two words, and they
#return the similarity between these two words.
#Lp and Lq denote the length of the 2 words. So Lk and Lv do.
#nSa stands for the count of same Chinese character at the same relative position
#in 2 words.
#nSi stands for the count of similar Chinese character in 2 words.
#Ssp is defined to characterize the adjustable space in a pair similarity
#measurement(1-10).

Input:
word pair(w_1, w_2)
Output:
$Simh = Hownet(w_1, w_2)$
$Simw = Word2vec(w_1, w_2)$
$Sim = (Simh + Simw)/2$
if $Simh > \alpha$ and $(Simh - Simw) > \beta$ **then**
    $Vsim1 = 2 * nSa/Lp + Lq$
    $Vsim2 = 2 * nSi/Lk + Lv$
    $Vsim = \sqrt{Vsim1 + Vsim2}$
    $Ssp = 10 - Sim$
    $Sim = Sim + Ssp * Vsim$
end if
return sim;

---

written documents and causal short texts. And these words also varied from the frequency of use and length, some of which have even more than one sense. Word pairs are constructed by an expert in computational linguistics, sometimes referring to Tongyici Cilin. And the final score is got by calculating the average score of twenty post-graduate students scores, who major in linguistics.

In addition, we choose a dump of Chinese wikipedia data, including Arts, History, Society, Technology, Entertainment and some other categories and select the most recent versions of all articles from 2004 to May 15, 20152. During preprocessing, phrase xml labels are transformed into the plain texts. We then remove all html labels and punctuations. On the other hand, we choose Sogou News as another corpus. These news data comes from several web sites in June–July 2012, which varies from the domestic, international, sports, social, entertainment and other 18 channels. Then we only retain the text. Then we use Python and the JieBa segmentation tool[2] to support Chinese word segmentation of sentences. There are three modes, we choose the accurate mode, for it cuts the sentence more accurate and suitable. Finally, we get 8,676,665 documents, 205,427,273 words from Chinese wikipedia data and 401,227,433 words from Sogou News in total [14].

---

[2] http://github.com/fxsjy/jieba.

3.2   Evaluation Method

Finally, we use Spearman(ρ) and Person(r) rank correlation coefficient, these two coefficients are widely used to evaluate the statistical dependence between our automatic computation results and the golden human labelled data.

Spearman correlation coefficient is shown below:

$$\rho = 1 - \frac{6\sum_{i=1}^{n}\left(R_{Xi} - R_{Yi}\right)^2}{n\left(n^2 - 1\right)} \tag{4}$$

where n is the number of observations, R_{Xi} and R_{Yi} are the standard deviations of the rank variables. The bigger r_R is, the more related computing results and labeled data are, and which means the results are better.

The Person correlation coefficient (r) is defined as:

$$r = \frac{\sum_{i=1}^{n}\left(X_i - \bar{X}\right)\left(Y_i - \bar{Y}\right)}{\sqrt{\left(X_i - \bar{X}\right)^2}\sqrt{\left(Y_i - \bar{Y}\right)^2}} \tag{5}$$

where X_i and Y_i are the raw score, $\bar{X}$ and $\bar{Y}$ are the mean value respectively, and n is the number of sample data.

3.3   Analysis

Base on the approach mentioned in previous section, we applied 10,000 pairs of words on it. Our approach consists of three parts, the hownet-based method, the word2vec-based method, the extreme-similar algorithm method and we denoted them by H, W, ES respectively. There has two models in the word2vec-based method, we use W-cbow and W-skip to respect them. Besides, we set $\alpha = 9$, $\beta = 7$, the values we choose depend on the follow experiment. Table 1 shows the results of 6 merging strategies trained with 2 groups of different corpora, where ρ represents the Spearman ρ between the result and the golden score, and r is the Pearson r between the result and the golden score [15].

As is shown in Table 1, the result differs when the usage of algorithm and corpora change. We can find that H or W are unable to get a good performance separately. This is because word2vec-based method (W) has two drawbacks in nature. Firstly, solely embedding technique cannot capture antonyms [16]. Secondly, there are some synonymy in Chinese vocabulary and the embedding method unable to calculate the similarity of pairs. And the hownet-based method (H) depends on the corpora, which used for train, and it is unable to calculate the similarity of unknown pairs of words. H + W improves the performance a lot. In this combination, they made up for each other's weaknesses. In addition, comparing the results of No. 1–3 and 8–10 we can see that H is more convincing. In terms of the quantity and quality of corpora, the larger scale does not absolutely mean the higher performance. Then comparing the results of No. 4, 5, No. 6, 7, No. 11, 12 and No. 13, 14, it illustrates the effectiveness of our extreme-similar algorithm. Finally No. 12 is selected as the best model.

Table 1. Results of 6 merging strategies based on different corpora

No.	Corpus	Model	ρ	r
1	News	H	0.241	0.311
2		W-cbow	0.285	0.290
3		W-skip	0.289	0.280
4		H + W-cbow	0.317	0.350
5		H + W-cbow + ES	**0.382**	**0.389**
6		H + W-skip	0.340	0.353
7		H + W-skip + ES	**0.409**	**0.409**
8	Wiki	H	0.241	0.311
9		W-cbow	0.282	0.279
10		W-skip	0.268	0.266
11		H + W-cbow	0.343	0.357
12		H + W-cbow + ES	**0.427**	**0.421**
13		H + W-skip	0.334	0.349
14		H + W-skip + ES	**0.413**	**0.415**

Figures 2 and 3 shows the influence of algorithm threshold (α and β). In this way, we choose the H + W-cbow + ES and Wiki to train the result. Firstly, we set $\alpha = 8$, observing the results after changing the value of β from 0 to 10. Next we set $\beta = 4$, then change the value of α from 0 to 10. We can easily see that, when the β is changeless, the result will have better performance as the α increasing, which indicates that it is more likely to believe that the word pair is extremely similar when the result of hownet-based method to reach a vary large level. It fits the fact that the higher artificial score is, the more similar the word pair is. In terms of β, it is obviously to say that when β is around 7, our experiment will perform better. Since we are more likely to believe the H, if the Simh is a lot larger than Simw, the average of two score are very likely to lower than the reality, so we must to refine the final score. And 7 is the best dividing line.

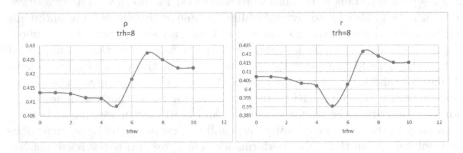

Fig. 2. The influence of β

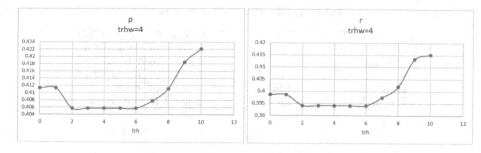

Fig. 3. The influence of α

4 Results

According to the two indicators in the above subsection, our best model is integrating the hownet-based method and word2vec-based method together, which is trained by wiki and using CBOW-model, with our extreme-similar algorithm ($\alpha = 9$, $\beta = 7$), in this part, we will show the great performance of this model. I will illustrate the effects of our model in three tables. We still use H, W-cbow and ES to express the same meaning, and Label is the score marked by people.

As is shown in Table 2. When the score of hownet-based method and the difference between hownet-based method to word2vec-based method both large, we add the value of hownet-based method and get our final result.

As is shown in Table 3, when both the score of hownet-based method and word2vec-based method show these two words are similar, we use our method to make them more approximately actual score.

In the last table, we will show the accuracy of our parameters. The first group we choose $\alpha = 8$, $\beta = 4$ (G1), the second group is our final model (G2) (Table 4).

Table 2. Part one of our results

w1	w2	H	W-cbow	H+W-cbow+ES	Label
安排	部署	10	2.368	**10**	9
入场券	门票	10	6.232	**8.11**	9
想法	主意	10	6.347	**8.174**	7.9
言语	语言	9.064	4.802	**6.933**	7
生命	性命	9.532	4.935	**7.234**	8.2

Table 3. Part two of our results

w1	w2	H	W-cbow	H+W-cbow+ES	Label
但	但是	10	8.80	**9.40**	9.3
跋扈	骄横	9.168	7.464	**8.315**	8.3
比如	譬如	10	6.871	**8.436**	8.7
会见	会晤	10	7.197	**8.599**	9.1

Table 4. Part three of our results

w1	w2	H	W-cbow	G1	G2	Label
信誉	信用	10	5.221	10	**7.6**	7.9
合同	协议	10	5.952	10	**7.976**	7.8
恼	怒	10	5.875	10	**6.743**	7.1
迂腐	墨守成规	9.167	4.32	10	**6.933**	7
言语	语言	9.064	4.802	10	**6.933**	7

So far, it is not difficult to find that we do improve the method to overcome the problem we proposed in the start.

5   Conclusion

Obviously, Chinese word similarity computation is a fundamental task for Chinese information processing, so we present a study for computing similarity of Chinese words. What is more, we devote our mind to the word pair which is extremely similarity, which others pay little attention.

Our strategy is to combine the hownet-based method, word2vector-based method and extreme-similar method. In our experiment, we train the word2vec-based method by using different models and corpora firstly. Then using our originality Extreme-Similar method, in this method, we utilize Tongyici Cilin to measure the similarity between 2 Chinese words in terms of the similar Chinese characters. At the same time, we record the sequence-similarity of two words. Combining them with Geometric Mean makes the score of this word pair closer to the larger one. After a series of operating, we finally get our best model, and the results are 0.427/0.421 of Spearman/Pearson, which outperforms the state-of-the-art performance to the best of our knowledge.

In the future, we are going to apply this method to a recommendation system. It helps us to clarify the relationship between words more accurately, which in turn assists us acquiring preferences from what people search or comment.

References

1. Mihalcea, R., Corley, C., Strapparava, C., et al.: Corpus-based and knowledge-based measures of text semantic similarity. In: AAAI, vol. 6, pp. 775–780 (2006)
2. Gan, M., Dou, X., Jiang, R.: From ontology to semantic similarity: calculation of ontology-based semantic similarity. Sci. World J. **2013**, 11 (2013)
3. Rothe, S., Schütze, H.: Autoextend: extending word embeddings to embeddings for synsets and lexemes. arXiv preprint arXiv:1507.01127 (2015)
4. Faruqui, M., Dodge, J., Jauhar, S.K., Dyer, C., Hovy, E., Smith, N.A.: Retrofitting word vectors to semantic lexicons. arXiv preprint arXiv:1411.4166 (2014)
5. Dai, L., Liu, B., Xia, Y., Wu, S.: Measuring semantic similarity between words using HowNet. In: 2008 International Conference on Computer Science and Information Technology, ICCSIT 2008, pp. 601–605. IEEE (2008)

6. Zhu, Y.-L., Min, J., Zhou, Y., Huang, X., Li-De, W.: Semantic orientation computing based on HowNet. J. Chin. Inf. Process. **20**(1), 14–20 (2006)
7. Liu, Q., Li, S.: Word similarity computing based on How-net. Comput. Linguist. Chin. Lang. Process. **7**(2), 59–76 (2002)
8. Xue, B., Fu, C., Shaobin, Z.: A study on sentiment computing and classification of sina weibo with word2vec. In: 2014 IEEE International Congress on Big Data (BigData Congress), pp. 358–363. IEEE (2014)
9. Levy, O., Goldberg, Y.: Neural word embedding as implicit matrix factorization. In: Advances in neural information processing systems, pp. 2177–2185 (2014)
10. Goldberg, Y., Levy, O.: word2vec Explained: deriving Mikolov et al'.s negative-sampling word-embedding method. arXiv preprint arXiv:1402.3722 (2014)
11. Levy, O., Goldberg, Y.: Dependency-based word embeddings. In: ACL (2), pp. 302–308. Citeseer (2014)
12. Tian, J., Zhao, W.: Words similarity algorithm based on Tongyici cilin in semantic web adaptive learning system. J. Jilin Univ. (Inf. Sci. Ed.) **28**(6), 602–608 (2010)
13. Wu, Y., Li, W.: Overview of the NLPCC-ICCPOL 2016 shared task: chinese word similarity measurement. In: Lin, C.-Y., Xue, N., Zhao, D., Huang, X., Feng, Y. (eds.) ICCPOL/NLPCC-2016. LNCS, vol. 10102, pp. 828–839. Springer, Cham (2016). doi:10.1007/978-3-319-50496-4_75
14. Fillmore, C.J., Wooters, C., Baker, C.F.: Building a large lexical databank which provides deep semantics. Citeseer (2001)
15. Rong, X.: word2vec parameter learning explained. arXiv preprint arXiv:1411.2738 (2014)
16. Ono, M., Miwa, M., Sasaki, Y.: Word embedding-based antonym detection using thesauri and distributional information. In: HLT-NAACL, pp. 984–989 (2015)

Measuring the Wellness Indices of the Elderly Using RFID Sensors Data in a Smart Nursing Home

Yuan Wu[1(✉)], Li Liu[2], Jinlong Kang[1], Lingling Li[1], and Bingqing Huang[3]

[1] School of Information Science and Engineering, Lanzhou University, Lanzhou 730000, China
hjf6318567@gmail.com
[2] School of Software Engineering, Chongqing University, Chongqing 400044, China
[3] The High School Attached to Northwest Normal University, Lanzhou 730070, China

Abstract. In this study, we present a low-cost, flexible and data-driven intelligent system which could monitor and determine the wellness conditions of the elderly living in a nursing home in relation to their daily activities. Changes of daily activities are obtained in real time for reasonable forecasting of wellness indices. These tasks are achieved by a framework integrating spatial and temporal contextual information for determining the wellness of the elderly. The daily activities of the elderly are detected through the location information collected by the Radio Frequency Identification (RFID) technology. A Support Vector Machine (SVM) model is trained using the activity data of 5 different elderly people living in a nursing home, and the results show that it performs well in forecasting wellness indices of the elderly.

Keywords: RFID · Smart nursing home · Wellness · SVM

1 Introduction

The number of elderly people in China is increasing, and this trend set to continue in future [1]. In the traditional opinion, the performance of daily activities of a normal elderly person seems to be steady and unchangeable, this suggests that a healthy elderly person is physically and mentally fit and leading a regular life. This concept suggests that indicators of daily activity measurements could determine the wellness of the elderly person. For example, the daily activities of an elderly person including basic behaviors like sleeping, reading books, walking, watching television, showering, taking dinner etc., are main indicators in determining the health status of the elderly [2]. However, the traditional assessments of daily activities are mostly done manually through interviews and questionnaires, and these procedures are time consuming and error prone [3], it could highly benefit from automatic assessment technologies.

Radio Frequency Identification (RFID), which uses electromagnetic fields to automatically identify and track tags attached to objects, is a fast developing technology for data collection and transmission, it could obtain data automatically and efficiently without human intervention [4]. RFID tags need not to be within line-of-sight of the readers to provide information in contrast with traditional tracking technologies like barcode. Moreover, RFID readers could simultaneously capture information of several

S. Song et al. (Eds.): APWeb-WAIM 2017 Workshops, LNCS 10612, pp. 66–73, 2017.
https://doi.org/10.1007/978-3-319-69781-9_7

tags. Because of the advantages of durability, small size and low cost, RFID tags can be easily embedded in the monitored objects and identified through computers or smart phones automatically.

In recent years, the number of the new family pattern called "4-2-1", which consists of 4 elderly people, a middle-aged couple and a child, is growing dramatically in China. In a "4-2-1" family, the middle-aged couple have the responsibility to take care of the elderly and the child, however, many young people choose to work and live in big cities while their parents stay in hometown, it is time consuming and costly for young people to care for their parents on their own, so living in a nursing home becomes a good choice for the elderly. Thus, an intelligent, low-cost and robust automatic engineering system is required to monitor and record daily activities and respond without delay when there is a change in the regular daily activity of an elderly person, so that necessary care could be delivered in time.

Elderly people living in a nursing home could be given better care with the information and communication technology. Thus, the requirement for an automatic system with intelligent mechanism for monitoring and recording daily activities of elderly is growing. Measurement of how well an elderly performs his/her daily activities could be implemented through analyzing the activity data. RFID technology has already invaded into healthcare and has attracted attention from a variety of research communities. Wicks and his colleagues has analyzed the potential benefits, implementation challenges and strategies of RFID applications in hospital [5]. [6] discusses some case studies and explored the important success factors for RFID technology adoptions in south-east Asian healthcare industry. [7] presents an academic review involving a wide range of RFID applications.

In this paper, we proposed an intelligent system based on RFID technology to monitor and record the location information of the elderly in a nursing home. The developed system is intelligent, robust and does not use any camera as it intrudes privacy. The system could collect location information of elderly people living in the nursing home. Healthcare specialists suggests that the best way to recognize health conditions of elderly before they become sick is to look for the changes in their daily activities [8–10]. Here we use SVM to build the health model with the collected location information of elderly to determine whether their health conditions are healthy or unhealthy. In the developed system, Human activity recognition and determination of the health condition as 'healthy' or 'unhealthy' through analyzing the collected location data are two important functions to be performed. Thus, health condition forecasting in a nursing home with RFID technology is a learning task, the main goal is to have the ability to recognize the change of the daily activities of the elderly through tracking the activities of the elderly in the nursing home.

The remainder of this paper is organized as follows: Sect. 2 describes the developed system and the implementation details. Section 3 presents the methods used to train the model and the data analysis results. Finally, Sect. 4 discusses conclusions and gives an outlook on future work.

2   System Description

There are two parts in the developed system: (i) location monitoring and recording module and (ii) intelligent health determination module. The first module is to record and collect the location data of the elderly in the nursing home, while the second module could perform data analysis and recognize changes of daily activities of the elderly. This architecture enables a system of an efficient RFID-based and data-driven health status forecasting system for a wide range of applications.

We conducted experiments and tested the robustness of the developed system in the department of geriatrics of the third people hospital in Lanzhou (Fig. 1), which is the most famous hospital specialized in mental disease in Gansu Province. The elderly living in the department of geriatrics all have mental diseases, they can't leave the department without doctors' permissions, so it could guarantee that the stability of the experimental scenario. There are bedrooms, activity room, dining room, reception room, bathroom and toilets in this department. RFID readers (Fig. 2(a)), which could capture RFID tag signals and transfer the information to the central computer through Wi-Fi network, and RFID stimulators (Fig. 2(b)), which stimulate RFID tags to transmit signals to RFID readers, were placed on the ceiling of each room and corridor. The operating frequency of the RFID reader we used is 433 MHz, while the operating frequency of the RFID stimulator is 125 kHz. The measuring range of RFID readers and stimulating range of RFID stimulator are both limited to 5 m in radius.

Fig. 1. The third people hospital in Lanzhou and the department of geriatrics

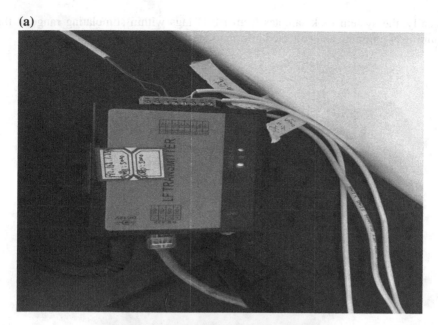

Fig. 2. (a) RFID reader used in the experiment, (b) RFID stimulator used in the experiment

The overall duration of our experiment was 20 days, 5 elderly patients (5 males) were chosen as the experimental subjects with the doctors', the patients' and the patients' family members' permissions. RFID tags, with unique identification for each elderly person, were embedded in their shoes or clothes, so that the tags will not influence their daily activities and those RFID tags were integrated into the developed system (Fig. 3).

Basically, the system took samples from RFID tags within stimulating range at the beginning of a minute and collected information of the elderly 24 h a day.

Fig. 3. Shoes and clothes used in the experiments

There are four parts in one location data (Fig. 4): (i) the object's identification; (ii) the sampling date; (iii) the sampling time; (iv) the identification of the RFID reader capturing signals from the RFID tag. In this experiment, the identification of the RFID reader was used to determine which room the elderly stays in at that one minute when his RFID tag signal was captured because the location of the RFID reader, where an object was recognized, could be considered as the coarse location of the recognized object. Figure 5 presented examples of the location data stored in the database.

TagID	DATE	TIME	READERID

Fig. 4. The format of the location information

	id	tagNum	date	time	ReaderID
☐	21651	82027157	2016-11-19	22:43:00	1
☐	21652	82027156	2016-11-19	22:43:00	1
☐	21653	52027158	2016-11-19	22:43:00	2
☐	21654	82028058	2016-11-19	22:43:00	2
☐	21655	82032353	2016-11-19	22:43:00	2

Fig. 5. Examples of the location data stored in the database

It produced 1440 location data for an elderly one day, if we used the original data to train the SVM model, the architecture of the model became extremely complicated and the time for training would become too long because of large amount of redundant data.

Thus, certain strategy would be adopted to process the original data. The department areas were divided into 5 types in this study (Fig. 6): bedroom, toilet (including bathroom and toilet), activity room (including activity room, reception room), dining room, and corridor, then we accumulated the time an elderly stayed in each type of areas of the department one day and used the time data as the input for SVM model.

Time in Bedroom	Time in Activity room	Time in Toilet	Time in Dining room	Time in Corridor

Fig. 6. The format of the activity information

The main goal of the study was to explore whether there is a link between daily activities of an elderly and his wellness index. Thus, we asked the doctors to assess the wellness conditions of these 5 elderly we selected every day and recorded them so that we could use them as labels in the supervised machine learning task.

3   Methods and Results

The SVM, presented by Vapnik and his colleagues, is a supervised learning algorithm based on VC (Vapnik-Chervonenkis) dimension theory and the structural risk minimization (SRM) principle [11]. It could be applied in both classification and regression. SVM has the ability to achieve the best compromise between the complexity and the learning ability of the model based on the limited sample information. In this study, we trained SVM model for classification, the basic principle of SVM for classification is to map the data into a high dimensional feature space through nonlinear mapping and to conduct a linear classification in this feature space [12].

In our experiment, input of the SVM model including time in bedroom, time in toilet, time in activity room, time in dining room and time in corridor on one day, while the doctors' assessments of the wellness indices of the elderly on that day, which were indicated as 'healthy' or 'unhealthy', were used as the labels, e.g. the output of the SVM model. The kernel function performed the non-linear mapping is radial basis function (RBF). Comparing with other kernel functions, RBF could reduce the computational complexity of the training procedure [13].

In our experiment, data of the elderly of the former 15 days, together with health assessments given by doctors of the former 15 days, were used to train the model, while the remaining data were used for testing the robustness of the model. The performance of the model was evaluated through identification accuracy. The identification accuracy was defined as:

$$accuracy = \frac{TPs + TNs}{TPs + TNs + FPs + FNs} \times 100\%$$

where TP indicates true positive, which means a positive sample is classified as a positive example; TN indicates true negative, which means a negative sample is classified as a negative example; FP indicates false positive, which means a negative sample is classified as a positive example; FN indicated false negative, which means a positive sample is classified as a negative example. The higher the classification accuracy is, the better performance the model obtains.

Figure 7 showed the accuracy of the classification for the testing data. The accuracy is 96%, the number of errors between actual health assessments of the elderly and predicted health conditions by SVM model is 1, which indicates that it is feasible to determine the wellness conditions of the elderly living in a stable environment like nursing home using this pattern.

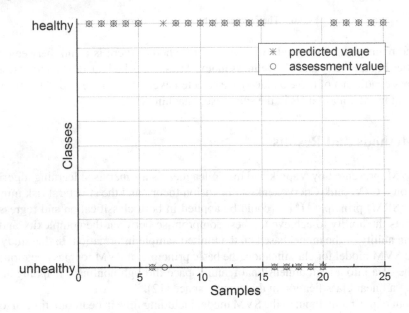

Fig. 7. Results of the experiment

4   Conclusions and the Future Work

In this paper, we developed an automatic engineering system to monitor and record location information of the elderly living in a nursing home, and used the data to determine the wellness indices of the elderly. An experiment on 5 elderly people during 20 days was conducted. The results suggested that it was feasible to predict the wellness conditions of the elderly using their activity data through SVM model, the accuracy of the prediction could reach 96%. It was indicated that time accumulation of different areas in a stable environment on one day could represent the activity information of the elderly to a certain extent, and the activity information could be used to predict wellness conditions of the elderly.

In the future, more parameters like body temperature could be combined with activity information for wellness indices forecasting of the elderly. We could ask the doctors give more detailed reports of the health conditions of the elderly and used them as labels to the supervised machine learning task. Category on the process of the original location data could also be changed, and we could explore whether there is a link between activity data and certain mental disease, like depression.

Acknowledgement. The authors would like to express their gratitude to all the subjects that participated in the experiments. This study is supported by Science and Technology Innovation Project of Foshan City, China (Grant No. 2015IT100095) and Science and Technology Planning Project of Guangdong Province, China (Grant No. 2016B010108002).

References

1. Wangzhou, Y., Suocheng, D., Youde, W., Renbo, L.: An analysis on the spatial distribution of population aging pressure in China. Chin. J. Popul. Resour. Environ. **10**(1), 122–128 (2012)
2. Suryadevara, N.K., Mukhopadhyay, S.C., Wang, R., Rayudu, R.K.: Forecasting the behavior of an elderly using wireless sensors data in a smart home. Eng. Appl. Artif. Intell. **26**(10), 2641–2652 (2013)
3. Philipose, M., Fishkin, K.P., Perkowitz, M., Patterson, D.J., Fox, D., Kautz, H., Hahnel, D.: Inferring activities from interactions with objects. IEEE Pervasive Comput. **3**(4), 50–57 (2004)
4. Yao, W., Chu, C.H., Li, Z.: The adoption and implementation of RFID technologies in healthcare: a literature review. J. Med. Syst. **36**(6), 3507–3525 (2012)
5. Wicks, A.M., Visich, J.K., Li, S.: Radio frequency identification applications in hospital environments. Hosp. Top. **84**(3), 3–9 (2006)
6. Vanany, I., Shaharoun, A.B. M.: Barriers and critical success factors towards RFID technology adoption in South-East Asian healthcare industry. In: Proceedings of the 9th Asia Pacific Industrial Engineering and Management Systems Conference, Bali, pp. 148–155 (2008)
7. Ngai, E.W.T., Moon, K.K.L., Riggins, F.J., Candace, Y.Y.: RFID research: an academic literature review (1995–2005) and future research directions. Int. J. Prod. Econ. **112**(2), 510–520 (2008)
8. Lawton, M.P., Brody, E.M.: Assessment of older people: self-maintaining and instrumental activities of daily living. Nurs. Res. **19**(3), 278 (1970)
9. Rogers, W.A., Meyer, B., Walker, N., Fisk, A.D.: Functional limitations to daily living tasks in the aged: a focus group analysis. Hum. Factors: J. Hum. Factors Ergonomics Soc. **40**(1), 111–125 (1998)
10. Tapia, E.M., Intille, S.S., Larson, K.: Activity recognition in the home using simple and ubiquitous sensors. In: Ferscha, A., Mattern, F. (eds.) Pervasive 2004. LNCS, vol. 3001, pp. 158–175. Springer, Heidelberg (2004). doi:10.1007/978-3-540-24646-6_10
11. Vapnik, V.: The support vector method of function estimation. In: Suykens, J.A.K., Vandewalle, J. (eds.) Nonlinear Modeling, pp. 55–85. Springer, Boston (1998). doi: 10.1007/978-1-4615-5703-6_3
12. Mukherjee, S., Osuna, E., Girosi, F.: Nonlinear prediction of chaotic time series using support vector machines. In: Proceedings of the 1997 IEEE Workshop on Neural Networks for Signal Processing VII, pp. 510–520. IEEE (1997)
13. Chauchard, F., Cogdill, R., Roussel, S., Roger, J.M., Bellon-Maurel, V.: Application of LS-SVM to non-linear phenomena in NIR spectroscopy: development of a robust and portable sensor for acidity prediction in grapes. Chemometr. Intell. Lab. Syst. **71**(2), 141–150 (2004)

Using Jaccard Distance Measure for Unsupervised Activity Recognition with Smartphone Accelerometers

Xianlong Wang[1], Yonggang Lu[1(✉)], Dachuan Wang[1], Li Liu[2(✉)], and Huiyu Zhou[3]

[1] School of Information Science and Engineering, Lanzhou University,
Lanzhou 730000, Gansu, China
1781368972@qq.com, ylu@lzu.edu.cn, 910379526@qq.com
[2] School of Software Engineering, Chongqing University, Chongqing 400044, China
dcsliuli@cqu.edu.cn
[3] Institute of Electronics, Communications and Information Technology,
Queens University of Belfast, Belfast BT3 9DT, UK
h.zhou@ecit.qub.ac.uk

Abstract. The rapid popularity of smartphones has led to a growing research interest in human activity recognition (HAR) with the mobile devices. Accelerometer is the most commonly used sensor of smartphone for HAR. Many supervised HAR methods have been developed. However, it is very difficult to collect the annotated or labeled training data for HAR. So, developing of effective unsupervised methods for HAR is very necessary. The accuracy of an unsupervised method, such as clustering, can be greatly affected by the similarity or distance measures, because the learning process of clustering method is completely depending on the similarity between objects. Although Euclidean distance measure is commonly used in unsupervised activity recognition, it is not suitable for measuring distance when the number of features is very large, which is usually the case in HAR. Jaccard distance is a distance measure based on mutual information theory and can better represent the differences between nonnegative feature vectors than Euclidean distance. It can also work well with a large number of features. In this work, the Jaccard distance measure is applied to HAR for the first time. In the experiments, the results of the Jaccard distance measure and the Euclidean distance measure are compared, using three different feature extraction methods which include time-domain, frequency-domain and mixed-domain feature extractions. To comprehensively analyze the experimental results, two different evaluation methods are used: (a) C-Index before clustering, (b) FM-index after using five different clustering methods which are Spectral Cluster, Single-Linkage, Ward-Linkage, Average-Linkage, and K-Medoids. Experiments show that, almost for every combination of the feature extraction methods and the evaluation methods, the Jaccard distance measure is consistently better than the Euclidean distance measure for unsupervised HAR.

Keywords: Activity recognition · Jaccard distance · Unsupervised HAR · Smartphone · Feature extraction

© Springer International Publishing AG 2017
S. Song et al. (Eds.): APWeb-WAIM 2017 Workshops, LNCS 10612, pp. 74–83, 2017.
https://doi.org/10.1007/978-3-319-69781-9_8

1   Introduction

With the improvement of health awareness of people and the continuous development of science technology, HAR is increasingly used in intelligent life, motion detection, and healthcare system [1–6]. The healthcare system can keep track of the health of user, and capture the activity of user for a long-term, which can help doctor to diagnose whether the user suffers from chronic diseases or not, and to urge the user to take proper exercise every day. The specialized medical equipment, wearable sensors or camera-based computer vision system have been used to recognize human activities, but they all require complex equipment and require camera to be placed in a fixed position, which is inconvenient for daily activity recognition.

Alternatively, smartphone is very popular now and people can carry it anytime, anywhere. Most smartphones are equipped with a rich set of embedded sensors. So smartphone has become an active field of research in the domain of perception and mobile computing. Within various sensors of smartphone, accelerometer is the most commonly used sensor for recording human activity signals. Khan et al. [7] have used a smartphone with a built-in triaxial accelerometer to collect five daily physical activities from five body positions, and the average classification accuracy of their method is about 96%. In the work of He and Li [8], three different sensors including accelerometer, gyroscope, and magnetic sensor embedded in a smartphone placed at the chest of a subject are used for HAR. The classification accuracy of their method is 95.03%. Their work also shows that within the three embedded sensors, the accelerometer is the most significant sensor for activity recognition.

Usually, raw time series data cannot be applied on HAR directly, feature extraction methods have to been used to produce a new data representation (called features) from the raw acceleration data before the classification. Popular features extracted from the acceleration data in the previous work can be divided into two types: time-domain features and frequency-domain features. Time-domain features include mean, variance, standard deviation, etc. [9], while frequency-domain features include frequency-domain entropy, discrete FFT coefficients, etc. [10]. In addition, mixed-domain features which include features from both time-domain and frequency-domain are also used in HAR [11].

In the traditional approaches to HAR, standard supervised classification methods, such as Support Vector Machine (SVM), K-Nearest Neighbor (KNN), Decision Tree, etc., have been used after feature extraction. Sun et al. [12] have used accelerometer embedded mobile phones to monitor seven common physical activities of seven subjects and have boosted the overall accuracy from 91.5% to 93.1% by improving the SVM method. Anjum and Ilyas [13] have collected a dataset of seven different activities, which include walking, running, ascending stairs, descending stairs, cycling, driving and remaining inactive using cell phone sensors and have evaluated a number of classification methods including Naive Bayes, Decision Tree, KNN and SVM. The Decision Tree classifier outperforms the other classifiers, and on average it produces a true positive rate of 95.2%.

All the above studies are using supervised classification methods to perform activity recognition. For the supervised methods, a large number of annotated data are required to train the classifiers. However, the data annotation is a difficult task which usually requires lots of time and efforts. So the unsupervised methods which can use the raw data directly for HAR have special advantages. In our previous work [14], an unsupervised method is proposed for HAR with time-domain features extracted from the data collected using smartphone accelerometers. It is shown that the unsupervised method is also effective to distinguish several daily activities. The unsupervised methods, such as clustering methods, usually classify data based on the similarities between the data points. So the accuracy of the unsupervised classification methods can be greatly affected by the distance measure (or similarity measure) applied on the feature vectors. Although the Euclidean distance measure is the most commonly used method, it may fail to measure the differences between features effectively in the high-dimensional space due to the curse of dimensionality. In many cases, the normalization of features is also needed for using the Euclidean distance measure. The drawbacks of the Euclidean distance measure can be avoided by using the Jaccard distance measure instead. Jaccard distance measures the degree of the overlap between two sets of nonnegative feature values, so the normalization of features is not needed. Jaccard distance not only considers the differences between the feature vectors, but also considers the absolute values of the feature vectors, so it can better represent the differences between nonnegative feature vectors than Euclidean distance [15]. Furthermore, because Jaccard distance is based on mutual information theory, it can be applied to measure the similarities between objects in high-dimensional space. In this work, the Jaccard distance measure is applied to HAR for the first time as far as we know, and the superiority of the Jaccard distance to the Euclidean distance is shown by the experiments.

2   Related Work

Recently, position and orientation independent activity recognition using smartphone accelerometers has also been investigated. Fan et al. [16] use resultant acceleration to eliminate the effects of the phone orientations. A feature set including seven time-domain features and three frequency-domain features is used for classifying five activities: staying still, walking, jogging, ascending stairs and descending stairs. The classification accuracy produced by their method is 80.29%. Miao et al. [9] have used accelerometer, gyroscope, proximity sensor, light sensor and magnetic sensor of a smartphone for HAR. The magnitude of linear acceleration combined with signals collected from gyroscope sensor and magnetic sensor are used. Then six time-domain statistical features are extracted for recognizing five typical physical activities which include staying still, walking, jogging, ascending stairs and descending stairs. Their results show the possibility of position and orientation independent HAR using smartphones without firm attachment to the body. In this work, orientation-independent HAR is realized by using the resultant acceleration data from the smartphones too.

Many clustering methods have been applied on unsupervised HAR. In the work of Machado et al. [17], the activities of standing, sitting, walking, running, lying are studied with mixed-domain features. Four clustering methods, including K-Means, Spectral Clustering, Mean Shift and Affinity Propagation (AP) based on Euclidean distance are used to distinguish different activities. In the subject-independent context, the accuracy rate of K-Means clustering reaches 88.75%, while the accuracy rate of other clustering algorithms ranges from 30% to 45%. In the case of subject-dependent context, the accuracy rate of K-Means clustering reaches 99.29% and the accuracy of other clustering methods ranges from 53% to 93.96%. In work of Gomes [18], K-Means, Mini Batch K-Means, AP, Mean Shift, DBSCAN, Spectral Clustering, and Ward-Linkage have been applied for HAR. They have found that the Spectral Clustering has produced the best results with an accuracy rate of 89.1 ± 8.8%. Amitha and Rajakumari [19] have proposed a system using Naive Bayes classifier combined with a hierarchical agglomerative clustering algorithm for activity tracking. In this work, Spectral Clustering, K-Medoids, and three commonly used hierarchical clustering methods, including Single-Linkage, Ward-Linkage, and Average-Linkage are used for HAR.

3 Method

3.1 Data Collection

In the experiments, both an UCI dataset and a dataset collected by ourselves are used for HAR. The UCI dataset includes triaxial acceleration data of walking, running, ascending stairs, descending stairs, sitting and standing of 30 subjects [20]. However, the data of 13 subjects have missing values or invalid values. So only the data of the rest 17 subjects are used in our experiments.

The dataset collected by ourselves are from 6 healthy volunteers with ages from 22 to 28. These activities are walking, jogging, ascending stairs, descending stairs, sitting and standing. The acceleration data with a sampling frequency of 50 Hz are collected.

3.2 Feature Extraction

Before feature extraction, to reduce the bias caused by sensor sensitivity and noise, a sliding window with 50% overlap is employed to divide the time series into smaller time windows. For the UCI dataset the window of 2.56 s is used, while for our dataset, the window of 5.12 s is used. Similar as previous works [21], the magnitude of the acceleration, A_{3a}, is used for feature extraction, which is insensitive to the orientation and position of the devices. A_{3a} can be represented as:

$$A_{3a} = \sqrt{A_x^2 + A_y^2 + A_z^2}$$

The magnitudes of the acceleration data within each sliding window are then used for feature extraction. Three feature extraction methods based on frequency-domain, time-domain and mixed-domain are used in the experiments, which are explained in details below:

(a) **Time-domain Features:** The same set of time-domain features used in the work of Miao et al. [9] are extracted, which includes Mean, Standard Deviation, Median, Skewness, Kurtosis, and Inter-Quartile-Range.

(b) **Mixed-domain Features:** The same mixed-domain feature set as reported in the
work of Figueira et al. [11] are extracted, which includes Root Mean Square (RMS),
Median Absolute Deviation (MAD), Standard Deviation, Spectral Roll On, Mean
Power Spectrum, Max power Spectrum.

(c) **Frequency-domain Features:** In the frequency-domain, to eliminate the effect of
edge samples in each window, a sine window is first multiplied to the data in the
window. The sine window is defined as:

$$win = \sin((i - 1)\pi/(winLen - 1))^2, \quad i \in [1{:}winLen]$$

where *winLen* represents the length of the window. Then FFT is applied to the data of
each window. In order to improve the frequency resolution and reduce spectral leakage,
the magnitudes of the Fourier coefficients are convolved with a hamming window of
size 5. Then the final results of each window are used as feature vectors.

3.3 Jaccard Distance Measure

In this work, Jaccard distance [15] is used to calculate the similarity between feature
vectors. If $X = (x_1, x_2, ..., x_n)$ and $Y = (y_1, y_2, ..., y_n)$ are two vectors and all x_i, y_i ($1 \le i \le n$) are non-negative, their Jaccard similarity coefficient is defined as:

$$J(X, Y) = \frac{\sum_{i=1}^{n} \min(x_i, y_i)}{\sum_{i=1}^{n} \max(x_i, y_i)},$$

and their Jaccard distance is defined as:

$$D_J(X, Y) = 1 - J(X, Y).$$

The value of the Jaccard distance ranges from 0 to 1, where 1 represents the identical
sets, while 0 represents the disjoint sets. Because in our work, the feature values extracted
using the three feature extraction methods are all greater than or equal to 0, Jaccard
distance can be readily applied.

4   Experiments

4.1   Design of the Experiments

To compare the effectiveness of the Euclidean distance and the Jaccard distance in
unsupervised HAR, the following procedures of the experiments are executed (as shown
in Fig. 1): (a) Collect the acceleration data; (b) Divide the time series data into small
time windows using a sliding window with 50% overlap; (c) Use the three feature
extraction methods; (d) Calculate the distance; (e) Use C-index to compare Jaccard

distance and Euclidean distance; (f) Cluster the data based on the distances between feature veetors and then measure the clustering results using FM-index.

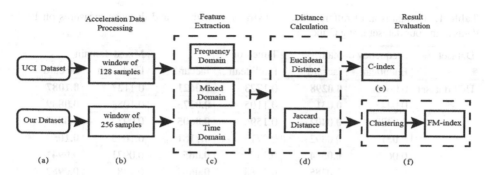

Fig. 1. Procedure of the experiments.

4.2   Evaluation Criterion

Both C-index [22] and FM-index [23] are used to evaluate the experimental results. C-index is used to measure the compatibility of the distance measure with the actual class labels. A smaller C-index value indicates a better compatibility where the data points have smaller distances within the same class and greater distances between different classes. The C-index is defined as:

$$C\_index = \frac{S - S_{min}}{S_{max} - S_{min}}$$

where S is the sum of the distances over all m pairs of objects from the same class, S_{min} is the sum of the m smallest distances if all pairs of objects are considered. S_{max} is the sum of the m largest distances out of all the pairs. The interval of the C-index values is between 0 and 1. A smaller C-index value indicates a better compatibility of the distance measure with the actual class labels.

The FM-index is used to evaluate the clustering results produced by a clustering method. The maximum value of the FM-index is 1 that means that the clustering result are the same as the actual result, while the minimum value is 0 that means the clustering result and actual result are completely different.

4.3   Experimental Results

For the UCI dataset containing 17 subjects and our dataset containing 6 subjects, the C-index is used to measure the distances generated by three different feature extraction methods and the two different distance measures. The results are shown in Table 1. From Table 1, it can be seen that, for 60 out of all 69 cases, the C-index produced using the Jaccard distance measure are better than that produced using the Euclidean distance measure, regardless of which feature extraction method is used. The results show that

Jaccard distance can measure the distances between feature vectors better than Euclidean distance.

Table 1. Comparison of different feature extraction methods and distance measures on UCI dataset and our dataset using C-index.

Dataset	Frequency domain		Time domain		Mixed domain	
	Euclidean	Jaccard	Euclidean	Jaccard	Euclidean	Jaccard
UCI dataset	0.0645	**0.0298**	0.1723	**0.0621**	0.1127	**0.1087**
	0.0445	**0.031**	0.1105	**0.0375**	0.1058	**0.0889**
	0.0439	**0.0402**	0.1596	**0.0605**	0.0932	**0.0865**
	0.031	0.0892	0.2715	**0.1181**	0.1093	**0.1053**
	0.0617	**0.0554**	0.147	**0.065**	0.0931	**0.0941**
	0.2034	**0.0285**	0.1488	**0.0691**	0.1383	**0.0987**
	0.0555	**0.0404**	0.2497	**0.0712**	0.1632	**0.1416**
	0.0606	**0.0281**	0.1274	**0.0438**	0.0608	**0.0568**
	0.0531	**0.0373**	0.2003	**0.0694**	0.0973	**0.0849**
	0.0406	**0.031**	0.1504	**0.0412**	0.1072	**0.0809**
	0.0577	0.0644	0.0789	**0.0408**	**0.0409**	0.0432
	0.0885	**0.0116**	0.0927	**0.0594**	0.0583	**0.0421**
	0.1229	0.1316	0.2175	**0.0849**	0.181	**0.1707**
	0.0608	**0.0311**	0.077	**0.0256**	0.0256	**0.0245**
	0.0586	**0.0254**	0.185	**0.0674**	0.1688	**0.1379**
	0.0745	**0.0299**	0.1779	**0.0857**	0.1356	**0.1279**
	0.0395	**0.0287**	0.1367	**0.0444**	**0.0581**	0.0624
Our dataset	0.0294	**0.0282**	0.1734	**0.0049**	0.0525	**0.0335**
	0.0314	**0.0303**	0.0803	**0.012**	0.0378	**0.0272**
	0.0296	**0.0239**	0.0401	**0.0047**	0.0219	**0.0169**
	0.027	**0.0253**	0.129	**0.0097**	0.0702	**0.065**
	0.0201	0.021	0.1652	**0.0155**	0.0606	**0.0421**
	0.0286	**0.0273**	0.4472	**0.0082**	0.0227	**0.0197**

Based on the distances computed using the two distance measures, five different distance-matrix based clustering methods which include Spectral Cluster, Single-Linkage, Ward-Linkage, Average-Linkage, and K-Medoids are used to cluster the dataset. The FM-index is then used to measure the clustering results. Table 2 shows the FM-indices produced using different combinations of the three feature extraction methods, the five clustering methods and the two distance measures on the UCI dataset and our dataset, where the FM-indices of all the subjects in each dataset are averaged.

It can be seen from Table 2, regardless of which combination of the feature extraction methods and clustering methods is used, the FM-indices produced using the Jaccard distance measure are consistently better than the results produced using the Euclidean distance measure. The results also show the superiority of the Jaccard distance measure over the Euclidean distance measure for distinguishing different activities. It can also be seen in Table 2 that the FM-indices produced using the frequency-domain feature

Table 2. Comparison of different feature extraction methods, distance measures and clustering methods using FM-index.

Dataset	Clustering method	Frequency domain		Time domain		Mixed domain	
		Euclidean	Jaccard	Euclidean	Jaccard	Euclidean	Jaccard
UCI dataset	Spectral clustering	0.543	**0.583**	0.344	**0.378**	0.387	**0.433**
	Single Linkage	0.807	**0.851**	0.723	**0.771**	0.612	**0.68**
	Ward Linkage	0.77	**0.81**	0.763	**0.791**	0.635	**0.664**
	Average Linkage	0.79	**0.871**	0.757	**0.814**	0.618	**0.687**
	K-Medoids	0.653	**0.654**	0.633	**0.645**	0.589	**0.604**
Our dataset	Spectral clustering	0.674	**0.688**	0.543	**0.582**	0.387	**0.442**
	Single Linkage	0.845	**0.895**	0.878	**0.904**	0.62	**0.658**
	Ward Linkage	0.895	**0.91**	0.887	**0.909**	0.652	**0.701**
	Average Linkage	0.869	**0.886**	0.836	**0.87**	0.636	**0.695**
	K-Medoids	0.749	**0.729**	0.739	**0.726**	0.62	**0.624**

extraction method are all better than these produced using the other two feature extraction methods, except the result of using Single-Linkage on our dataset. This indicates that the frequency-domain features are more effective than the other two features for unsupervised HAR.

To further compare the Jaccard distance measure and the Euclidean distance measure for activity recognition, the FM-indices produced using the five clustering methods are averaged for each combination of the distance measure and the feature extraction

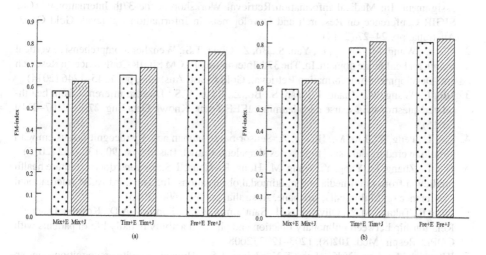

Fig. 2. Comparison of different combinations of the feature extraction methods and the distance measures using the average of FM-index of the five clustering methods on (a) UCI dataset, and (b) our dataset, where "Fre", "Tim", "Mix" represent frequency-domain, time-domain, mixed-domain feature extraction methods respectively, and "E", "J" represent Euclidean distance and Jaccard distance respectively.

methods. The results are shown in Fig. 2. It can be seen that the FM-index produced using Jaccard distance is better than that produced using Euclidean distance regardless of which feature extraction method is used, while the FM-index produced using the frequency-domain feature extraction method is better than that produced using the other two feature extraction methods.

5   Conclusion

In this paper, the Jaccard distance measure is proposed to replace the Euclidean distance measure for unsupervised HAR. Both the C-index before clustering and the FM-index after clustering show the superiority of the Jaccard distance measure over the Euclidean distance measure for unsupervised HAR. It is also found that the frequency-domain feature extraction method is better than the time-domain feature extraction method and the mixed-domain feature extraction method for unsupervised HAR using the proposed method. Future work includes applying the Jaccard distance measure on more datasets and improving the feature extraction methods.

Acknowledgments.   This work is supported by the National Science Foundation of China (Grants No. 61272213) and the Fundamental Research Funds for the Central Universities (Grants No. lzujbky-2016-k07).

References

1. Nie, L., Akbari, M., Li, T., Chua, TS.: A joint local-global approach for medical terminology assignment. In: Medical Information Retrieval Workshop at the 37th International ACM SIGIR Conference on Research and Development in Information Retrieval, Gold Coast, Australia, pp. 24–27 (2014)
2. Nie, L., Wang, M., Zhang, L., Yan, S., Bo, Z., Chua, T.S.: Wenzher: comprehensive vertical search for healthcare domain. In: The 37th International ACM SIGIR Conference on Research and Development in Information Retrieval, Gold Coast, Australia, pp. 1245–1246 (2014)
3. Nie, L., Wang, M., Zhang, L., Yan, S., Bo, Z., Chua, T.S.: Disease inference from health-related questions via sparse deep learning. IEEE Trans. Knowl. Data Eng. **27**(8), 2107–2119 (2015)
4. Liu, L., Peng, Y., Liu, M., Huang, Z.: Sensor-based human activity recognition system with a multilayered model using time series shapelets. Knowl. Based Syst. **90**, 138–152 (2015)
5. Nie, L., Zhang, L., Yang, Y., Wang, M., Hong, R., Chua, T.S.: Beyond doctors: future health prediction from multimedia and multimodal observations. In: The 23rd ACM International Conference on Multimedia, Brisbane, Australia, pp. 591–600 (2015)
6. Pitta, F., Takaki, M.Y., Oliveira, N.H., Sant'Anna, T.J.P., Fontana, A.D., Kovelis, D., et al.: Relationship between pulmonary function and physical activity in daily life in patients with COPD. Respir. Med. **102**(8), 1203–1207 (2008)
7. Khan, A.M., Lee, Y.K., Lee, S.Y., Kim, T.S.: Human activity recognition via an accelerometer-enabled-smartphone using kernel discriminant analysis. In: The 5th International Conference on Future Information Technology (FutureTech), Busan, Korea (South), pp. 1–6 (2010)

8. He, Y., Li, Y.: Physical activity recognition utilizing the built-in kinematic sensors of a smartphone. Int. J. Distrib. Sens. Netw. **9**(4), 481580 (2013)
9. Miao, F., He, Y., Liu, J., Li, Y., Ayoola, I.: Identify typical physical activity on smartphone with varying positions and orientations. Biomed. Eng. Online **14**(1), 32 (2015)
10. Huynh, T., Schiele, B.: Analyzing features for activity recognition. In: Proceedings of the 2005 Joint Conference on Smart Objects and Ambient Intelligence: Innovative Context-Aware Services: Usages and Technologies, Grenoble, France, pp. 159–163 (2005)
11. Figueira, C.R., Matias, R., Hugo, G.: Body location independent activity monitoring. In: The 9th International Joint Conference on Biomedical Engineering Systems and Technologies (BIOSTEC), Rome, Italy, pp. 190–197 (2016)
12. Sun, L., Zhang, D., Li, B., Guo, B., Li, S.: Activity recognition on an accelerometer embedded mobile phone with varying positions and orientations. In: Yu, Z., Liscano, R., Chen, G., Zhang, D., Zhou, X. (eds.) UIC 2010. LNCS, vol. 6406, pp. 548–562. Springer, Heidelberg (2010). doi:10.1007/978-3-642-16355-5_42
13. Anjum, A., Ilyas, M.U.: Activity recognition using smartphone sensors. In: 10th Consumer Communications and Networking Conference (CCNC), Las Vegas, USA, pp. 914–919 (2013)
14. Lu, Y., Wei, Y., Liu, L., Zhong, J., Sun, L., Liu, Y.: Towards unsupervised physical activity recognition using smartphone accelerometers. Multimedia Tools Appl. **76**, 1–19 (2016)
15. Levandowsky, M., Winter, D.: Distance between sets. Nature **234**(5), 34–35 (1971)
16. Fan, L., Wang, J., Wang, H.: Activity recognition based on position-independent smartphone. J. Comput. Inf. Syst. **10**(11), 4921–4928 (2014)
17. Machado, I.P., Gomes, A.L., Gamboa, H., Paixão, V., Costa, R.M.: Human activity data discovery from triaxial accelerometer sensor: non-supervised learning sensitivity to feature extraction parameterization. Inf. Process. Manag. **51**(2), 204–214 (2015)
18. Gomes, A.L.G.N.: Human activity recognition with accelerometry: novel time and frequency. Doctoral dissertation, Champalimaud Neuroscience Programme (2014)
19. Amitha, R., Rajakumari, K.: Hierarchical clustering based activity tracking system in a smart environment. Int. J. Comput. Sci. Tech. **3**, 2229–4333 (2012)
20. UCI dataset. https://archive.ics.uci.edu/ml/datasets/Human+Activity+Recognition+Using+Smartphones
21. Wang, D., Liu, L., Wang, X., Lu, Y.: A novel feature extraction method on activity recognition using smartphone. In: Song, S., Tong, Y. (eds.) WAIM 2016. LNCS, vol. 9998, pp. 67–76. Springer, Cham (2016). doi:10.1007/978-3-319-47121-1_6
22. Hubert, L., Schultz, J.: Quadratic assignment as a general data-analysis strategy. Br. J. Math. Stat. Psychol. **29**(2), 190–241 (1976)
23. Fowlkes, E.B., Mallows, C.: A method for comparing two hierarchical clusterings. J. Am. Stat. Assoc. **78**(383), 553–569 (1983)

Speech Emotion Recognition Using Multiple Classifiers

Kunxia Wang[1]([⊠]), Zongcheng Chu[1], Kai Wang[1], Tongqing Yu[1], and Li Liu[2]([⊠])

[1] Key Lab of Artificial Architecture, Anhui Jianzhu University, Hefei, China
kxwang@ahjzu.edu.cn, joechu_ml@163.com, wangkai@keking.cn,
18326693328@163.com
[2] School of Software Engineering, Chongqing University, Chongqing, China
dcsliuli@cqu.edu.cn

Abstract. The research topic of how to automatically identify the emotional state of speakers received much attention. In this paper, we mainly focus on speech emotion recognition and develop an audio-based classification framework for identifying five different emotions in our audio database where the audio segments are from Chinese TV plays. First, acoustic features were extracted from the audio segments using Wavelet analysis, then feature selection is implemented based on Information gain and Sequential Forward Selection in the purpose of reducing irrelevant information as well as dimension reduction. Our classification framework is constructed over three base classifiers: SVM, Adaboost and Randomforest. Considering of the fact that a single classifier is in the limitation of recognition capability, decision fusion methods are applied to aggregate different prediction labels. According to the experiment on our database, the fusion methods we proposed show better performance.

Keywords: Emotion recognition · Multiple classifier · Decision fusion

1 Introduction

As a major part of pattern recognition, speech emotion recognition has attracted much interests [1], since it is indispensable to human computer interactions [2].

During the process of speech emotion recognition, the classification process is quite an important step to connect high-level speech emotion with voice signals. So far, there are a variety of classification techniques used in the field of speech emotion recognition including Gaussian Mixture Model (GMM), Artificial Neutral Networks (ANN), Hidden Markov Model (HMM), Support Vector Machine (SVM), etc.

Traditionally, the classification process is based on single classifier. However, one single classifier has certain restrictions, thus may not be able to get the most desirable results. Mainstream classification algorithm are SVM, Random Forest [3], Adaboost [4], etc. SVM has low generalization error rate but it is sensitive to parameter tuning and kernel function selection. Random Forest can process high dimensional data and possess the ability to generalize, however it

© Springer International Publishing AG 2017
S. Song et al. (Eds.): APWeb-WAIM 2017 Workshops, LNCS 10612, pp. 84–93, 2017.
https://doi.org/10.1007/978-3-319-69781-9_9

has the problem of overfitting especially when the data noise is relatively large. Adaboost owns the ability to avoid overfitting and attain high accuracy while it needs to consume large amount of time to train.

In this paper, we use Wavelet analysis as our major approach for acoustic features extraction. The tasks we are facing are to assign a single emotion label to the audio clip from five emotions: Anger, Disgust, Happy Sad and Neutral. So, the classification methods we adopt will have direct impact on the final results. Then we proposed fusion methods of classifiers for speech emotion recognition. Both majority vote strategy and weighted vote are adopted. Based on the prediction results of these classifiers, we can further perfect the accuracy.

The remainder of the paper is organized as follow: Sect. 2 introduces the related literature. Section 3 briefly describes the models we use. In Sect. 4, we proposed two methods for decision fusion. In Sect. 5, we conduct our experiment based on EESDB database. Then the final conclusion is given in Sect. 6.

2   Related Work

In the literature of speech emotion recognition, previous works has made much effort to explore methods to achieve high-quality feature extraction and get accurate classification results.

As for audio features, the AVEC 2011–2015 challenge employed the open source software openSMILE [5] to extract LLDs features which include the Energy and spectral related LLDs. Tufekci and Gowdy [6] propose a new feature vector consisting of mel-frequency discrete wavelet coefficients (MFDWC). The MFDWC are obtained by applying the discrete wavelet transform (DWT) to the mel-scaled log filterbank energies of a speech frame. Dharanipragada and Rao describes a robust feature extraction method [7] for continuous speech recognition. Central to the method is the minimum variance distortionless response (MVDR) method of spectrum estimation and a feature trajectory smoothing technique for reducing the variance in the feature vectors.

For data classification, Suykens and Vandewalle [8] discuss a least squares version for support vector machine (SVM) classifiers. Svetnik et al. [9] introduce and investigate Randomforest for predicting a compound's quantitative or categorical biological activity based on a quantitative description of the compound's molecular structure. In their work, they built predictive models for six cheminformatics data sets and their analysis demonstrates that Random Forest is a powerful tool capable of delivering performance that is among the most accurate methods to date. Bergsta et al. [10] apply Adaboost to select from a set of audio features that have been extracted from segmented audio. Their classifier proved to be the most effective method for genre classification at a contest in music information extraction.

For fusion strategies of the results getting from different classifiers, Sun et al. [11] propose a novel hierarchical classification framework, which combines the feature-level and decision-level fusion strategy for all of the extracted multimodal features. Kuncheva et al. [12] present a simple rule for adapting the class combiner to the application. Decision templates (one per class) are estimated with the same training set that is used for the set of classifiers. Then these templates are matched to the decision profile of new incoming objects by some similarity measure.

3 Classifiers for Emotion Recognition

Considering that one single classifier is not of high reliability to identify five various emotions, we proposed three different classification techniques which are SVM, Adaboost and Randomforest to have better prediction of our final results. The reason we choose these classifiers is that they are of powerful classification capability in our pre-testing and the optimal parameters will be discussed later.

3.1 SVM

The SVM method is able to transform the sample space into a high dimension or even a infinite dimension (Hilbert space) using a mapping function which is known as kernel function. So, in this case, it is possible to solve a Non-Linear problem in a higher dimensional space. The most common kernel functions are RBF Kernel and Gaussian Kernel.

In our experiment, we tried both RBF and Gaussian Kernel and finally adopted Gaussian Kernel as mapping function. Then, the grid search method is used for parameter adjustment to find out the optimal ones.

3.2 Adaboost

Adaboost is an iterative algorithm, its core idea is to apply different weak classifiers on the same training set.Then these weak classifiers together to form a stronger final classifier (strong classifier). Basically, Adaboost algorithm follows three steps.

First, the weight distribution of training data is initialized. Each training sample is given the same weight at the beginning.

$$D_1 = (w_{11}, w_{12}, ..., w_{1N}), w_{1i} = \frac{1}{N}, i = 1, 2, ..., N \tag{1}$$

Second, iterative process is performed.Basically, we follow three steps.

(A) Using the training data set with weight distribution D_m , the basic classifier is obtained.

$$G_m(x) : \chi \rightarrow -1, +1 \tag{2}$$

where denotes the weight distribution in iterative round.

(B). Calculating the classification error rate of on the training data set.

$$e_m = P(G_m(x) \neq y_i) = \sum_{i=1}^{N} w_{mi} I(G_m(x) \neq y_i) \tag{3}$$

(C) The weight of the basic classifier in the final classifier is obtained.

$$\alpha_m = \frac{1}{2} log \frac{1 - e_m}{e_m} \tag{4}$$

Third, combining weak classifiers so as to ensemble a stronger classifier.

$$f(x) = \sum_{m=1}^{M} \alpha_m G_m(x) \tag{5}$$

$$G(x) = sign(f(x)) = sign(\sum_{m=1}^{M} \alpha_m G_m(x)) \tag{6}$$

The basic estimator for Adaboost is a Decision tree and the parameter max_depth is set to 2. There are overall 800 Decision trees as our basic estimators.

3.3 Randomforest

Randomforest is one of ensemble learning methods. It consists of successive decision trees which are relatively dependent from each other. So, asynchronous calculation pattern is possible in a Randomforest. By using bootstrap sample of the training set, we can obtain k different decision trees. Compared with the traditional node splitting strategy using all the variables, in a Randomforest, each node is split by using a subset of all variables. Finally, the different prediction results produced by k decision trees are aggregated using the strategy of majority voting. It turns out to have good performance on classification problem. Moreover, it has a good quality of against overfitting problem.

4 Fusion Method

We explored two types of fusion methods in this paper: majority vote and weighted vote.

Majority vote is the simplest method which assigns the sample to class that had the most votes. The biggest weakness is that all the classifiers have the same authority, with no thought for their respective abilities in our specific classification task. Furthermore, simple weighted vote (SWV) [13] is proposed. The decision weight of each classifier is determined according to the classification accuracy on testing set. So, the weights can be calculated as.

$$w_k = \frac{a_k}{\sum_i a_i} \tag{7}$$

After the weights are obtained, the class with the highest score is our final prediction. The specific calculation process is shown as.

$$\hat{c}(x) = argmax \sum_k w_k \delta(\hat{c}_k(x_i), c_j) \tag{8}$$

$$\delta(a, b) = \begin{cases} 1, a=b \\ 0, otherwise \end{cases} \tag{9}$$

where $\hat{c}(x)$ is the prediction of sample x_i.

5   Experiment Result

Our experiment was based on Chinese elderly emotion database (EESDB) [14] which contains overall 783 audio samples extracted from 22 speakers (11 female and 11 male) from Chinese TV plays. There are five main emotion categories including Happy, Angry, Disgust, Neutral and Sad. Specifically, the database shows uneven distribution in which the Angry category has 261 samples while Disgust has only 53. Since the data is quite unbalanced, marco precision is used as our metrics for accuracy evaluation. Then, we extract acoustic features from the audio segments using Wavelet analysis [15, 16]. To exclude irrelevant information as well as to reduce the dimension, we implement feature selection by using the methods of information gain [17,18] and sequential forward selection [19,20] which show improvement in our classification results. Figure 1 clearly shows the whole framework of our work.

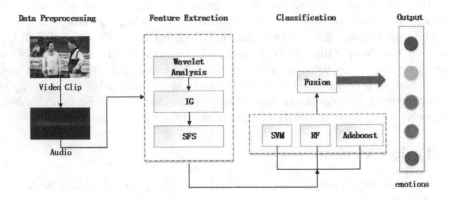

Fig. 1. Structure drawing of speech emotion recognition system

We first extracted acoustic features from audio segments using Wavelet analysis. A total of 5760 features were extracted for emotion recognition. First, we use wavelet package extraction to acquire a six dimension wavelet package coefficient, then calculating maximum, minimum, mean, median and ample variance of each dimension. After the first order difference calculation and second order difference calculation, we acquired a 90 dimension feature set, named improved wavelet package coefficient. Totally, We conduct 64 grades wavelet package extraction and get a 5760 dimension feature set to represent an audio.

On the basis of previous work, we process the feature selection based on the theory of information gain. We calculate the information gain of each feature, and according to the threshold, we reserve features which can meet the requirements. The experimental setting threshold starts at 2.1144 with size of 0.0032. Under each threshold, we get corresponding feature set and recognition rate. By using this method, we successfully select 2535 features from 5760 features, which are equally representative as the raw feature set, while effectively achieve dimension reduction.

Then, we apply the theory of sequential forward selection and choose SVM as the classifier. Initial feature subset include just one IG-selected improved Wavelet package coefficient. Next, we continue adding new ones to the subsets. Whenever the recognition rate decreases, the latest added one will be deleted. By implementing this process, we optimized the feature function and obtained a 1279 dimension feature.

Holdout method is presented in our classification task. The whole data set is separated into two sets, called the training set and the validation set. We set the ratio of training set to validation set is seven to three. We tried all the three classifiers for ten times cross validation (training set and validation set are randomly separated each time but the ration remains the same). The results are shown in Fig. 2.

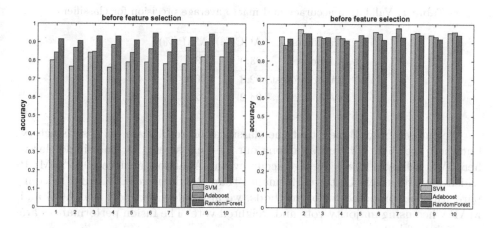

Fig. 2. Structure drawing of speech emotion recognition system

As we can see in Fig. 2, feature selection did improve the prediction accuracy as a whole, although Randomforest shows slight decline in accuracy. Both support vector machine and Adaboost show higher accuracy. So, we can draw the conclusion that the feature selection work is effective and it contributes to our results.

More specifically, in order to know more details about our classification results. Both marco average precision (MAP) and accuracy (ACC) are presented in this paper to evaluate the performance of different classifiers. These two measurements are given as below.

$$MAP = \frac{1}{s} \times \sum_{i=1}^{s} P_i \tag{10}$$

$$P_i = \frac{TP_i}{TP_i + FP_i} \tag{11}$$

$$accurancy = \frac{\sum_{i=1}^{s} TP_i}{\sum_{i=1}^{s} (TP_i + FP_i)} \tag{12}$$

where s denotes the number of overall emotion categories. P_i is the precision of the emotion category. and represent the true positive predictions and false positive predictions of the emotion category. represents the precision of emotion category.

Table 1 shows the distinctive ACC and MAP of different classification methods. The results are based on averages of ten times cross validation experiments.

Table 1. Validation accuracy and macro average precision for classifiers

Classifiers	ACC(%)	MAP(%)
Adaboost	93.9	93.2
Randomforest	92.8	90.1
SVM	94.2	92.4

We can see from Table 1 that the classification results are quite satisfactory but there are slight differences among our base classifiers. For one single classifier, SVM achieved 92.8% average accuracy and Adaboost achieved 94.2% MAP which have relatively better performance than Randomforest. In order to compare with our proposed fusion methods, we also calculated both the ACC and MAP when using majority vote and weighted vote. The fusion performances are shown in Table 2.

Table 2. Fusion performance on the validation set

	ACC(%)	MAP(%)
Majority vote	94.9	2.3
Weighted vote	94.2	94.6

As we can see from Table 2, the average accuracies of these two vote strategies are slightly higher than the average of one single classifier which means our fusion methods have positive effects on the final result, but the MAP still remain the same level.

In addition. Table 3 clearly presents how each classifiers recognition rate on specific emotions. The category of Neutral has the highest recognition rate among five emotions while the category of Sad has the lowest. The confusion matrix with the best results are shown in Table 4. We noticed that Angry has the most labels

Table 3. Recognition rate on specific emotion

	Angry	Disgust	Happy	Neutral	Sad
Adaboost	95.2	98.5	90.8	99.3	87.2
Randomforest	98.8	84.8	88.9	99.6	78.3
SVM	98.8	87.0	88.1	98.4	89.7

Table 4. Confusion matrix of the best prediction result on validation set

	Angry	Disgust	Happy	Neutral	Sad
Angry	83	0	0	0	0
Disgust	2	9	0	1	0
Happy	0	0	32	1	0
Neutral	0	0	0	66	0
Sad	0	0	0	8	33

while Disgust has only nine labels. Considering of the unbalanced distribution of our training set (261 samples for Angry, 53 for Disgust, 122 for Happy, 222 for Neutral and 125 for Sad), we have reason to believe that it will have strong impact on our final prediction results. Since the majority of the samples belong to Angry, the prediction results will be inclined to this category.

Besides of the distinctive predictive capability on different emotions presented in Table 3, we also noticed that during our experiment process, running time greatly varied among three models. To be more specific, Adaboost takes much longer time than the rest of two. SVM is the fastest model. We conducted experiment to test both the running time and Memory usage of each model as shown in Table 5. The running statuses of each model are obtained using python module time and memory _ profiler. Memory_ profiler can clearly show the memory usage increment of each line of code. The item of Memory usage shown in Table 5 is the total Memory usage of model training and prediction process. As we can see in Table 5, Adaboost is a time-consuming model, its running time is 20 times longer than Randomforest, 317 times than SVM. Randomforest takes up much more Memory space than SVM and Adaboost. On the whole, the average accuracy of each base classifier is quite close to each other, but there exist

Table 5. Comparision of classifiers and optimal parameters

	Run time	Memory (Mib)	Avg ACC(%)	g	c	Max depth	N estimator
Adaboost	153.6	0.680	93.9	–	–	2	1000
Randomforest	7.52	9.718	92.8	–	–	800	–
SVM	0.484	0.023	94.2	0.0039	16	–	–

huge differences on running time and Memory usage. SVM is a great model for our classification task, both the running time and Memory usage are the lowest. Meanwhile, it has the highest accuracy among three classifiers. So, SVM is the optimal classifier in this task. The parameter tuning is based on grid search method and the optimal parameter for each classifier is presented in Table 5.

6   Conclusion

In this paper, we proposed a new classification framework for identifying different emotions from audio segments. Having extracted acoustic features from raw audio data using Wavelet analysis, we applied feature selection methods including information gain and sequential forward selection to the original high-dimension acoustic features, finally remained overall 1280 features. Then three different classifiers were utilized for classification task. Moreover, two types of decision-level fusion methods were presented in our work, namely majority vote and weighted vote. Our proposed framework was evaluated on Chinese elderly emotion database and it turns out to have good performance. In future work, we will try more classification models and seek a more advanced fusion method to aggregate the prediction results.

Acknowledgements. This work was supported by the Open Project Program of the National Laboratory of Pattern Recognition (NLPR) (NO. 201700014), Anhui Provincial Natural Science Foundation (No. 1708085MF167), and Anhui Prov-ince Key Laboratory project of affective computing and advanced intelligent machines under grant ACAIM160103. Any correspondence should be made to Li Liu and Kunxia Wang.

References

1. El Ayadi, M., Kamel, M.S., Karray, F.: Survey on speech emotion recognition: features, classification schemes, and databases. Pattern Recogn. **44**(3), 572–587 (2011)
2. Fragopanagos, N., Taylor, J.G.: Emotion recognition in human–computer interaction. Neural Netw. **18**(4), 389–405 (2005)
3. Liaw, A., Wiener, M.: Classification and regression by randomForest. R News **2**(3), 18–22 (2002)
4. Zhu, J., Zou, H., Rosset, S., et al.: Multi-class adaboost. Stat. Interface **2**(3), 349–360 (2009)
5. Eyben, F., Weninger, F., Gross, F., Schuller, B.: Recent developments in opensmile, the Munich open-source multimedia feature extractor. In: Proceedings of the 21st ACM International Conference on Multimedia, pp. 835–838. ACM, October 2013
6. Tufekci, Z., Gowdy, J.N.: Feature extraction using discrete wavelet transform for speech recognition. In: Proceedings of the IEEE Southeastcon 2000, pp. 116–123. IEEE (2000)
7. Dharanipragada, S., Rao, B.D.: MVDR based feature extraction for robust speech recognition. In: Proceedings of 2001 IEEE International Conference on Acoustics, Speech, and Signal Processing (ICASSP 2001), vol. 1, pp. 309–312. IEEE (2001)

8. Suykens, J.A.K., Vandewalle, J.: Least squares support vector machine classifiers. Neural Process. Lett. **9**(3), 293–300 (1999)
9. Svetnik, V., Liaw, A., Tong, C., et al.: Random forest: a classification and regression tool for compound classification and QSAR modeling. J. Chem. Inf. Comput. Sci. **43**(6), 1947–1958 (2003)
10. Bergstra, J., Casagrande, N., Erhan, D., et al.: Aggregate features and AdaBoost for music classification. Mach. Learn. **65**(2–3), 473–484 (2006)
11. Sun, B., Li, L., Wu, X., et al.: Combining feature-level and decision-level fusion in a hierarchical classifier for emotion recognition in the wild. J. Multimodal User Interfaces **10**(2), 125–137 (2016)
12. Kuncheva, L.I., Bezdek, J.C., Duin, R.P.W.: Decision templates for multiple classifier fusion: an experimental comparison. Pattern Recogn. **34**(2), 299–314 (2001)
13. Moreno-Seco, F., Iñesta, J.M., de León, P.J.P., Micó, L.: Comparison of classifier fusion methods for classification in pattern recognition tasks. In: Yeung, D.-Y., Kwok, J.T., Fred, A., Roli, F., de Ridder, D. (eds.) SSPR /SPR 2006. LNCS, vol. 4109, pp. 705–713. Springer, Heidelberg (2006). doi:10.1007/11815921_77
14. Wang, K.X., Zhang, Q.L., Liao, S.Y.: A database of elderly emotional speech. In: Proceedings of International Symposium on Signal Processing, Biomedical Engineering Information, pp. 549–553 (2014)
15. Wang, K., An, N., Li, L.: Speech emotion recognition based on wavelet packet coefficient model. In: 2014 9th International Symposium on Chinese Spoken Language Processing (ISCSLP), pp. 478–482. IEEE, September 2014
16. Daubechies, I., Sweldens, W.: Factoring wavelet transforms into lifting steps. J. Fourier Anal. Appl. **4**(3), 247–269 (1998)
17. Ververidis, D., Kotropoulos, C.: Fast and accurate sequential floating forward feature selection with the Bayes classifier applied to speech emotion recognition. Signal Process. **88**(12), 2956–2970 (2008)
18. Tao, Y., Wang, K., Yang, J., An, N., Li, L.: Harmony search for feature selection in speech emotion recognition. In: International Conference on Affective Computing and Intelligent Interaction (ACII), pp. 362–367. IEEE, September 2015
19. Jin, Y., Song, P., Zheng, W., Zhao, L.: A feature selection and feature fusion combination method for speaker-independent speech emotion recognition. In: IEEE International Conference on Acoustics, Speech and Signal Processing (ICASSP), pp. 4808–4812. IEEE, May 2014
20. Kächele, M., Zharkov, D., Meudt, S., Schwenker, F.: Prosodic, spectral and voice quality feature selection using a long-term stopping criterion for audio-based emotion recognition. In: 22nd International Conference on Pattern Recognition (ICPR), pp. 803–808. IEEE, August 2014

Intelligent Taxi Dispatching Based on Improved Artificial Fish Swarm Algorithm

Zhiwei Luo[1], Rong Xie[2(✉)], Wangyi Huang[2], and Yiwei Shan[2]

[1] School of Automation, Huazhong University of Science and Technology,
1037 Luoyu Road, Wuhan 430074, China
zhiweiluo@hust.edu.cn

[2] International School of Software, Wuhan University, 39 Luoyu Road, Wuhan 430079, China
{xierong,wangyihuang,shanyiwei}@whu.edu.cn

Abstract. Taxi plays an important role in public transportation. However, in real life, it usually takes a long time for passengers to wait for an unoccupied taxi, while number of no-load taxi is very high. Optimized scheme is required to handle global taxi dispatching. The paper presents a novel method of taxi dispatching based on Artificial Fish Swarm Algorithm (AFSA). To avoid slow convergence and falling into local optimum, standard AFSA is improved using strategies of optimizing visual range, step size and swimming behavior. We compare our improved algorithm with standard AFSA and Particle Swarm Optimization through the evaluation with real data of GPS trajectory of 12,000 taxis in Beijing and make performance analysis of effects of parameter configuration to dispatching algorithm. The experimental results show good effectiveness of our method.

Keywords: Taxi · Taxi dispatching · Improved Artificial Fish Swarm Algorithm (AFSA) · Intelligent Transportation

1 Introduction

Nowadays, taxis provide people with fast and comfortable service. However, there are still some problems in taxi dispatching. Lack of a good solution to global taxi dispatching, which bringing about asymmetric information between taxi drivers and passengers, it usually costs a long time for passenger to wait for an unoccupied taxi; while simultaneously taxi drivers wander in streets, which leads to severe waste of fuel and traffic resources. In recent years, under the help of software for taxi reservation, for example, "DiDi" in China, unoccupied taxi nearby passenger can be quickly assigned to passenger who calls for a taxi by mobile phone. Though this method indeed reaches some achievements in alleviating the contradiction between passenger and taxi driver, it fails to make global optimization on taxi resources. Efficiently taxi dispatching is urgently required in dealing with this issue.

Methods for taxi dispatching are studied in the past years, which are mainly divided into two stages, manual scheduling and intelligent scheduling. With burgeon of taxis amount, due to low effectiveness, manual scheduling cannot meet passenger's demand.

Intelligent taxi scheduling shows promise. Gilbert et al. pioneerly improved taxi dispatching method by using global positioning system (GPS). In their simulation model with 16 teams and 300 cars in each team, the time wasted on taxi dispatching was decreased by 50–60%. Rosenschein and Zlotkin [1] proposed a method with association structure to allocate taxi for passengers. Aiming at minimum time waste on taxing arriving at passenger's position, Lee et al. [2] proposed intelligent taxi scheduling system based on road traffic information. Dorer and Calisti [3] proposed a method based on dynamic traffic information for the optimization for taxis' driving route. Seow et al. [4, 5] proposed the idea of distributed taxi dispatching. Sung and jong [6] predicted cost for transportation on each road for next period of time based on real-time and historical GPS data, and then used optimized Floyd-Washa algorithm for minimized path length. Sen et al. [7] proposed a path selection algorithm for time-varying traffic networks based on minimum expectations and variance. For practical applications, US Module Company once developed a vehicle dispatch management system which could achieve real-time GPS positioning, monitored the current state of taxis, and sent cars in a reasonable manner. In the recent, some researches tend to develop a recommender system for searching for passengers and vacant Taxis [8].

Differing from traditional methods mentioned above, learning from the thought of swarm intelligence in artificial intelligence, this paper presents a novel method to deal with the issues of taxi dispatching through AFSA. Optimization is made by dispatching taxis to an area with high density of passengers calling for taxis to realize the maximum efficiency of traffic resource utilization and minimum of waiting time.

2 AFSA and Its Improvements

2.1 Artificial Fish Vision Model

Under natural environment, as we can observe, fishes tend to go some areas with rich food resources. Such phenomenon can be interpreted with some behaviors as follows. (1) Foraging behavior. Fish usually swims in water without a certain target. But once food is found, fish will quickly swim to the food. (2) Clustering behavior. Individual fish tends to join with some other fishes to become swarm without certain guidance or organization but gather together in water to gain strengthened acuity to predators and resistance to disasters. (3) Following behavior. When a fish finds food in some areas, other fishes nearby would immediately change direction towards their companions. (4) Random behavior. If fish does not perform one of the above behaviors, it will search for food randomly in water.

In order to simulate these characteristics of natural fish, we can build an artificial fish vision model as follows. $X = (X_1, X_2, ..., X_n)$ is the current state of artificial fish, where X_i ($i = 1, 2, ..., n$) is each variable to be optimized. Food concentration of the current location of X is $Y = f(X)$, representing the value of objective function. $vision$ is the current visual range of artificial fish. X_v is location of artificial fish's viewpoint at a certain moment. $step$ is the maximum movement length for artificial fish's one step. $X_{l next}$ is the next position after artificial fish makes a movement, which can be calculated according to Eq. (1).

$$X_{inext} = X_i + rand() \times step \times (X_i - X_j)/\left\|X_i - X_j\right\| \tag{1}$$

where, $rand()$ generates a random number between [0, 1]. X_i and X_j represent the state of two companion artificial fishes respectively, and distance between them can be calculated as $d_{ij} = \|X_i - X_j\|$.

2.2 Standard AFSA Improvement

Although standard AFSA [9] has many advantages, some shortcomings can be found in actual applications. Neshat et al. [10] once made a survey of the existing artificial fish swarm algorithm. In the paper, we make improvements on it in the aspects of optimizing visual range and step size, as well as swimming behaviors as follows.

Improvement 1: Visual range and step size. Vision range *vision* has a great influence on the algorithm for both behavior and convergence performance. In general, scale of visual range will impact convergence and search ability. With rather larger visual range, artificial fish's behavior mainly embodied in following behavior and clustering behavior, and the algorithm will have quick convergence and strong global search ability. Adversely, with rather smaller visual range, artificial fish's behavior mainly embodied in foraging behavior and movement pattern show more randomness, and algorithm will be with slow convergence and cost more computing resources, but has strong local search ability. On the other hand, step size *step* will also impact on algorithm convergence. When step size is small, increment of step size will accelerate convergence of algorithm, but when it exceeds a certain value, the increase of step size will reduce convergence, and in the even worse case, leading to oscillation and failure to find optimal solutions. So, at the primary iterations of the algorithm, rather larger scale of visual range and step size should be chosen in order to accelerate convergence and enhance search ability of global optimal result. When algorithm gradually converge, the given result is the optimal solution, step length and visual range should be adjusted to rather small size to avoid oscillation and make sure that the result will approach the optimal solution progressively. The specific parameter configuration can be determined by Eq. (2).

$$\begin{cases} a = exp(-30 \times (t \,/\, t_{\max})^s) \\ vision = vision \times a + vision_{min} \\ step = step \times a + step_{min} \end{cases} \tag{2}$$

where, t is the current number of iteration. t_{max} is the maximum number of iteration. The minimum value of visual field is $visual_{min} = 0.001$. $step_{min} = 0.0002$. s is an integer which is greater than 1, between [1, 30] in the paper. Using the method, vision range and step size are the maximum at the beginning, and then gradually become smaller and finally reach the minimum.

Improvement 2: Swimming behaviors. In the standard AFSA, when artificial fish tries *trynum* times, if it still does not satisfy with forwarding condition, it will perform random behavior. However, if artificial fish has been already in the vicinity of the optimal solution, it may swim far away from the optimal position which will affect the results of the algorithm. In the implementation of cluster behavior and following behavior, if the algorithm has approached to global optimal solution, then forging behavior may cause fish to be deviated from the global optimal solution. The same problem will be encountered for performing clustering behavior and following behavior. Therefore, the improvements on swimming behaviors can be performed in the following aspects. (1) Forging behavior. When artificial fish has tried *trynum* times, if food concentration in the current state is greater than $1/n$ of optimal food concentration Y_{best}, then current state is maintained; otherwise random behavior is performed. (2) Clustering behavior. Clustering behavior will be continued only if food concentration Y_c at center of all companion fishes found in the vision region is greater than $1/n$ of optimal food concentration Y_{best}; otherwise, foraging behavior is performed. (3) Following behavior. The following behavior will be continued only if the maximum food concentration Y_{max} of all companion fishes found in vision range is greater than $1/n$ of optimum food concentration max; otherwise foraging behavior is performed.

3 Taxi Dispatching

3.1 Taxi Dispatching Problem (TDP)

The basic idea of taxi dispatching is to dispatch taxi to an area with high density of passengers. Similarly, AFSA makes artificial fishes search for an area with highest concentration of food. Taxi dispatching model can be described as follows. The scope that taxi dispatching can be divided into R areas. Suppose there are P passengers calling for taxi in this scope at a moment, with amount of passengers P_i in area i ($i = 1, 2, ..., R$), then $P = \sum_{i=1}^{R} P_i$. The amount of empty taxis in each area is represented by M_i. Suppose that a taxi T_j can accept the maximum no-load distance being d_{T_j} with the above conditions. TDP, that taxi dispatching is to make taxis go to the area with highest density and also no-load distance for taxi should be as short as possible, is descried as $max\ Opportunity(P_i, T_j) \wedge (Distance(P_i, T_j) \leq d_{unloading\_threshold})$. The possible dispatching results may be as follows. (1) The area where passengers' density is the largest will be satisfied first, and followed by the area with the second largest density of passengers, and so on. (2) Assuming that there are N areas with a relatively larger density, number of passenger of every taxi's currently area is P_i. Crowding factor cf is used to determine whether number of taxi gathered in a certain area exceeds a certain threshold. Set the area with the maximum density of passengers to N_{max}, number of passenger in this area to P_{max}, and number of free taxi in range to M_{max}. So, if $P_{max}/M_{max} > CP_i$, then the taxi will be distributed to N_{max}.

3.2 Taxi Dispatching Model

The waste of taxi resource is mainly caused by inappropriate dispatching. A feasible solution is to quickly dispatch available taxi to passenger nearby according to dynamic passenger's order. With terminal platform on taxi and client of mobile communication by passenger's mobile phone, passengers' information, such as location and time, can be captured from client and uploaded to server. Then they will be used as feedback for conducting proper guidance for taxi drivers. In this way, reasonable dispatching of taxi resources can be performed according to dynamic information.

Applying AFSA to taxi dispatching, we can regard passengers calling for taxi as food resource in a virtual pond, and taxi as an artificial fish changing its behaviors according to change of environment. Let X_i be location of taxi's visual point at a certain time, $step$ be the maximum step of taxi to move at one time, and X_{inext} be location of taxi after one step forward. Distance between two taxis X_i and X_j can be calculated by $d_{ij} = \|X_i - X_j\|$. Then movement of taxi can be modeled as Eq. (3).

$$X_{inext} = X_i + rand() \times step \times \left(X_c - X_i\right)/\|X_c - X_i\| \tag{3}$$

In real-life environment, passengers usually gather around some hotspot area, such as shopping malls, schools, railway stations and etc. So, in some certain scope, there may simultaneously exist several areas with high passenger density. When simulating this situation, area with different density of passengers is abstracted as a spatial local maximum point, where there is a unique global optimal maximum point. Finally, through performing ASFA, artificial fish distributed randomly will be dispatched to the final position. We can compare number of artificial fish near the global maximum point and the average distance of taxis from the optimal location, and use their final location to determine the algorithm's performance and effect to solve the problem.

3.3 Implementation

The steps of taxi dispatching based on our improved AFSA are described as follows.

Step 1: Initialize setting of the algorithm, including initial number of taxi, random distribution of initial position of taxi, location of passenger, visual range $vision$, length of one step forward $step$, crowding factor cf etc.
Step 2: Calculate size of passengers' density at location of all initial locations of taxis and record them in an array of the optimal state ($bestx_i$, $bexty_i$).
Step 3: Implement evaluation function, the results of it will be used for taxi to choose one of the four basic behaviors of artificial fish, i.e. foraging behavior, clustering behavior, following behavior, and random behavior. The optimal behavior will be selected to implement.
Step 4: Update optimal state array ($bestx_i$, $bexty_i$).
Step 5: Check whether termination condition is satisfied or not. If termination condition is satisfied, output the result; otherwise go back to execute Step 3.

4   Experiment and Performance Test

4.1   Raw Data and Test Function

The original data used in our experiment came from GPS big data which was generated by 12,000 taxis in Beijing in November 2012.

This experiment is performed by MATLAB R2013a in Windows 7 system. Experimental parameters are initial number of artificial fish *fishnum*, vision range of artificial fish *vision*, step size of one step forward *step*, passenger's crowding factor *cf*, number of attempt *trynum* and maximum number of iteration *maxgen*.

The essence of TDP is how to optimize allocation of taxi resources to regional centres with the highest density of passengers through an effective dispatching method. The purpose of this experiment is to test whether artificial fishes can be eventually clustered in the region with the largest concentration of food using the improved AFSA, i.e. in a certain spatial range, centre of region with global optimal maximum. Simulation test function, i.e. $z(x, y) = \sin(x) \times \sin(y)/(x \times y)$, is selected in our experiment.

As shown in Fig. 1, the test function, having a unique global maximum, with the size of 1 located at (0, 0), and many local optimal extremes near the global optimal maximum point, can be used to test effectively whether the results of the algorithm are in globally optimal feature or not. In addition, as standard AFSA or other optimization algorithms is often easy to fall into local optimal or shock near the local optimal region, but the improved AFSA through theoretical analysis that the searching results can be well gathered around global optimal maximum point, this test function is more suitable for verifying optimization performance of the algorithm in the experiment. In the simulation experiment, the initial position of taxi is randomly distributed at a position that maybe far from the global maximum point. After executing the improved AFSA, we can compare the result of the optimization algorithm with the result of the simulation function, and to check whether the distribution results can be clustered near the global optimal maximum point.

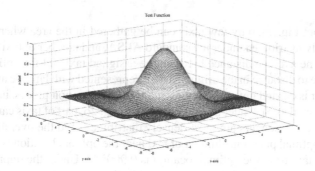

Fig. 1. Simulation of test function.

4.2   Comparison Analysis

Figure 2 gives comparison results of distribution map of taxi dispatching among standard AFSA, Particle Swarm Optimization (PSO) and improved AFSA. Black represents distribution-intensive area of passengers, red asterisk as the initial distribution of taxi, and blue circle as the final distribution of taxis using the algorithms. The horizontal coordinate indicates distance x between taxi and the area with the maximum density of passengers, and the vertical coordinate indicates distance y between taxi and the area with the maximum density of passengers. Figure 2 is a partial enlarged view of central area, number of taxi is *fishnum* = 216.

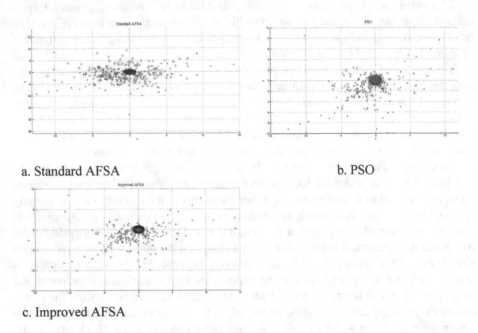

a. Standard AFSA   b. PSO

c. Improved AFSA

Fig. 2. Comparison of taxi dispatch distribution map among standard AFSA, PSO and improved AFSA.

As seen from Fig. 2a, only few taxis can be gathered in the area where passengers are most densely distributed under the standard AFSA, and number of taxi near optimal location is 2. The average distance between the assigned taxi and the optimal location is 8.33346 (Due to distance in the original data is represented with latitude and longitude, this experiment is carried out at a certain scale for magnification processing, so there is no specific physical unit). Under the PSO, shown in Fig. 2b, part of taxis can be gathered in the area where passengers are most densely distributed, and the overall distribution is close to the optimal position. Number of taxi near the optimal location is 123, average distance of the taxi from the optimal location is 0.900788. Under the improved AFSA, shown in Fig. 2c, most of taxis can be gathered to the area where passengers are most densely distributed, with number of taxi near the optimal location is 183, the average distance of taxi from the optimal location is 0.820658. Through the experiment, the improved AFSA represents its better optimization performance and better optimization effect comparing with standard AFSA and PSO.

4.3 Performance Analysis

We also analyze the influence of parameter setting on optimization performance of our improved algorithm as follows.

The impact of number of attempt *trynum* of taxi on optimal performance of the algorithm is shown in Fig. 3. As we can see, *trynum* has little effect on the optimization performance, if other parameters are set to the same. When *trynum* increases, artificial fish can search for more locations, but there is no guarantee that it can find a better location.

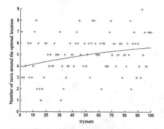
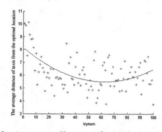

a. Number of taxi around the optimal location   b. Average distance from the optimal location

Fig. 3. Influence of *trynum* on optimal performance of the improved algorithm.

The effect of moving step *step* of taxi (artificial fish) on optimization performance of the algorithm is shown in Fig. 4. Under the condition that the other parameters are set to the same, the effect of improved AFSA is very good when *step* is about 50–65% of *vision*. Within a certain region of view, the greater the *step* is, the faster is to reach the area with the optimal maximum point, and the more potential position will be found when random behavior is performed. However, when *step* exceeds a certain range, it causes artificial fish to shock in the vicinity of optimal value, that optimization performance of the algorithm will be greatly weakened.

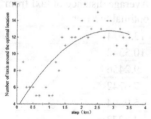
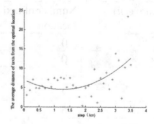

a. Number of taxi near the optimal location   b. Average distance from the optimal location

Fig. 4. Influence of *step* on optimal performance of the improved algorithm.

The influence of crowding factor *cf* on the optimization result of the algorithm is shown in Fig. 5. Under the other parameters are set to the same, the effect of *cf* on the experimental results is obvious when it is small. The greater *cf* is, the greater number of taxi near the optimal position is, and the smaller the average distance of taxi from the optimal position is. It is shown that the introduction of cf has obvious effect on the optimization experiment and is better to avoid the local optimum. For the case of cf is large, the effect on number of taxi near the optimal location and on the average distance of taxi from the optimal position are both less. This is because clustering behavior and following behavior are all optimized. Fish will implement following behavior only if concentration of food at the center of all companion fishes (or the maximum food concentration of all companion fishes) is greater than $1/n$ of optimal food concentration Y_{max}; otherwise foraging behavior is performed.

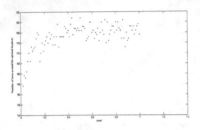 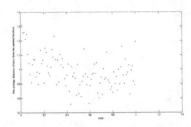

a. Number of taxi near optimal location   b. Average distance of taxi from optimal location

Fig. 5. The influence of cf on optimal performance of the improved algorithm.

The influence of *maxgen* on optimal performance of the algorithm is shown in Table 1. Under the other parameters are set to the same, when *maxgen* reaches 1,500 and above, there is no great difference in the final results. It is proved that when *vision*, *step*, *maxgen* and *cf* is fixed, we need to adjust *maxgen* to about 4000 in order to assure fast convergence of the algorithm and avoidance of shocking in the vicinity of all optimal maximum area.

Table 1. Influence of *maxgen* on optimal performance of the improved algorithm.

Iteration times (*maxgen*)	Number of taxi near optimal location	Average distance of taxi from optimal location
5	0	13.3674
10	0	10.5342
50	0	9.2456
100	2	7.6742
500	5	5.9894
1000	7	4.8273
1500	9	3.5927
4000	14	1.4283
5000	12	2.5856
10000	8	4.4769

5 Conclusion

In the paper, we propose a novel solution to taxi dispatching on the basis of improving the standard AFSA in the aspects of an adaptive visual field and step size, as well as swimming behaviors to optimize behavior of artificial fish. Some simulation experiments are conducted using MATLAB, showing that our improved AFSA has a better performance on taxi dispatching than standard AFSA and PSO.

In the paper, our experiment stays at the level of MATLAB simulation. Further work shall be combined with map to achieve more convincing experimental results.

Acknowledgment. This work is supported by Suzhou Science and Technology Bureau - International Cooperation Programme of Science and Technology under grant no. SH201213. The authors would like to thank Wei Pan, Maohui Lu and Yingfei Xiang at Wuhan University for their helpful algorithm development.

References

1. Rosenschein, J.S., Zlotkin, G.: Rules of Encounter: Designing Conventions for Automated Negotiation Among Computers, pp. 25–46. MIT Press, Cambridge (1994)
2. Lee, D.H., Wang, H., Cheu, R.L., et al.: A taxi dispatch system based on current demands and real-time traffic information. Transp. Res. Rec. **1882**, 193–200 (2004)
3. Dorer, K., Calisti, M.: An adaptive solution to dynamic transport optimization. In: Proceedings of the 4th International Joint Conference on Autonomous Agents and Multiagent Systems, Industry Track, Utrecht, The Netherlands, pp. 45–51 (2005)
4. Seow, K.T., Dang, N.H., Lee, D.H.: Towards an automated multiagent taxi-dispatch system. In: IEEE International Conference on Automation Science and Engineering, Scottsdale, AZ, USA, 22–25 September 2007, IEEE Xplore, pp. 1045–1050 (2007)
5. Seow, K.T., Dang, N.H., Lee, D.H.: A collaborative multiagent taxi-dispatch system. IEEE Trans. Autom. Sci. Eng. **7**(3), 607–616 (2010)
6. Sung, S.K., Jong, H.L.: A study on design of dynamic route guidance system using forecasted travel time based on GPS data and modified shortest path algorithm. In: IEEE/IEEJ/JSAI International Conference on Intelligent Transportation Systems, Tokyo, Japan, IEEE Xplore, pp. 44–48 (2002)
7. Sen, S., Pillai, R., Joshi, S., et al.: A mean-variance model for route guidance in advanced traveler information systems. J. Transp. Sci. **35**(1), 37–49 (2001)
8. Yuan, J., Zheng, Y., Zhang, L., et al.: Where to find my next passenger. In: Proceedings of the 13th International conference on Ubiquitous computing (UbiComp 2011), New York, pp. 1–10. ACM (2011)
9. Li, X.: A new intelligent optimization method-Artificial fish school algorithm. Hangzhou: Zhejiang University (2003)
10. Neshat, M., Sepidnam, G., Sargolzaei, M., et al.: Artificial fish swarm algorithm: a survey of the state-of-the-art, hybridization, combinatorial and indicative applications. Artif. Intell. Rev. **42**(4), 965–997 (2014)

HotSpatial 2017

Correction of Telecom Localization Errors by Context Knowledge

Minmin Zhu[1], Buyang Cao[1(✉)], Mingxuan Yuan[2], and Jia Zeng[2]

[1] School of Software Engineering, Tongji University, Shanghai, China
buyang60@hotmail.com
[2] Huawei Noah's Ark Lab, Shatin, Hong Kong

Abstract. Telecom localization that had aroused widespread atten-
tions of major telecommunication operators has become vital in recent
years. However, current available technologies suffer from high localization errors, typically with mean errors more than 100 m. In order to tackle
this problem, in this paper we leverage context knowledge to reduce the
localization error. To this end, we propose a framework adopting several modified filter methods in terms of context to eliminate localization
errors that cannot be easily detected by the existing localization algorithms. We apply the optimized filter methods combining with the context knowledge to verify the effectiveness of our methodologies according
to the experiments based on the telecom localization utilizing the GPS-
associated MR data in the downtown area of Shanghai, China.

1 Introduction

LBS requires relatively accurate locations of mobile phone users. Nevertheless,
the traditional telecom positioning approaches suffer from either low precision
(e.g., the range-based methods have the typical mean errors by hundreds of
meters) or high costs (the fingerprinting methods have to maintain a fingerprint
database). The recent mea- surement report (MR)-based positioning systems
instead have many advantages includ- ing availability in most mobile phones
and being active whenever users make phone calls and use mobile broadband
services. It has been considered as a very useful com- plement to GPS. However,
MR positioning systems still cannot achieve high precision.

In the view of the current domestic and international research status, localization schemes on Wireless Sensor Network (WSN) data can be classified as three
categories, (1) range-based methods, (2) fingerprinting methods, and (3) model-
based methods. The range-based methods use range measurements as physical
models [6], which record the TOA (Time of Arrival) and AOA (Angle of Arrival),
TDOA (Time Difference of Arrival) and RSSI (Radio Signal Strength Indicator)
of transmitting wireless signal by the unknown node hardware receiving from
external symbol node, then transform these distance metric values to the distance upon which the related algorithms such as trilateration and triangulation
method, maximum likelihood estimation method can be employed.

© Springer International Publishing AG 2017
S. Song et al. (Eds.): APWeb-WAIM 2017 Workshops, LNCS 10612, pp. 107–117, 2017.
https://doi.org/10.1007/978-3-319-69781-9_11

Fingerprint positioning algorithm is a feature matching algorithm, which uses a plurality of signal strength values of wireless routers in positioning environment, and establish the off-line fingerprint database [9] by collecting and training, then to match the real-time collected fingerprints in the positioning process and fingerprint database in order to estimate the best matching positions. The last method is a model-based localization, which uses related machining learning algorithms to learn our excited position estimation, such as Random Forest (RF) algorithm and Artificial Neural Network (ANN) algorithm. This method serves the real GPS position as our training label to build our localization model by training the Measurement Report(MR) provided by mobile phone service providers.

Despite the localization methods we studied had achieved good position estimation results, we had found that lots of localization errors having large deviations from the real values, which could mainly reflect in either the oscillations of position in the same road section or deviating far away from the original real road segments. The direct use of the localization algorithm to calculate the location of a user may produce big error due to the change in user position. Furthermore, the user movement is not very smooth and it might be affecting the real-time positioning system performance and stability seriously. Inevitably, there will be a lot of noise during the process of the signal transmission in addition to the noise produced by the localization algorithm itself, which is a vital reason causes the error of localization.

Furthermore, we found that in addition to the estimated positions deviating seriously from the true positions, there are serious velocity variations in the predicted locations. We believe the speeds between two adjacent positions of a user trajectory do not appear reasonable according our daily life experience. Therefore, solving these localization error problems become very important to improve localization accuracy as the basis for LBS.

In this paper, we introduce several filtering algorithms to remove the abnormal location coordinates created by the positioning algorithm and to further improve the positioning accuracy of the positioning system built on model-based localization.

Our study indicates that it is necessary to propose a practical and effective data postprocessing method to resolve a variety of positioning errors. In our view, every GPS point is not isolated position but contextual related because these GPS points represent real physical locations of user trajectory and normally it could predict the next position according the current or last position, and/or infer the last position according the current position. As we discussed above, the work we attempt to accomplish in this paper is to design a sliding window filter based on map-matching algorithm, which mainly combines with two items of the big rich context knowledge from the entire user trajectory and road network data, in order to eliminate the influence of errors in our localization models.

To summarize, we make the following contributions in this paper:

- Novel application of several filters in trajectory data processing and separated from data preprocessing work

- The first to introduce the algorithm of using context knowledge based on map-matching
- The combination of mathematical filter algorithm model and spatial-temporal data.

The rest of this paper is organized as follows, Sect. 2 describes the problem of localization error to be solved. Section 3 introduces the theory for proposed model and error correction algorithm. Section 4 presents the related experiment results. Finally, Sect. 5 concludes with the efficiency of the proposed method and the possible future work.

2 Background

When mobile users make phone calls using mobile phones or use mobile broadband services, their phones connect to telecom networks, e.g., GSM. The network next generates measurement report (MR) data. The MR data records the received signal strength indicator (RSSI) of nearby base stations to support communication services, etc. On the other hand, the widely-used location based services (LBSs) have accumulated lots of over-the-top (OTT) global positioning system (GPS) data in telco networks. We then use the GPS data as the training labels to learn accurate MR-based positioning systems. Figure 1 shows the data flow of an MR-based positioning system. LBSs generate low sampling OTT GPS locations (green dots). With the OTT GPS locations as label data, the MR-based positioning system can train the high sampling MRs by using machine learning models. Since the GPS locations are numeric data, we can adopt the classic machine learning algorithm named Random Forests to solve a regression problem. When the training model is ready, given the MR records without labels, we predict the GPS locations (yellow dots) with respect to such MR records. In this way, with the predicted GPS points, we can fully recover the entire trajectory.

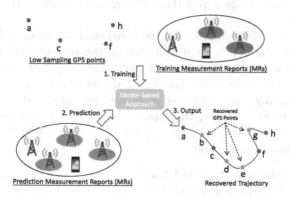

Fig. 1. Model-based MR positioning systems (Color figure online)

Suppose that the above MR records have ground truth (i.e., the GPS locations), we can measure the positioning precision by comparing the predicted location to the ground truth GPS location. The previous work [10] can achieve a mean error of around 80 m. Though the positioning precision obtained by applying the RF algorithm is much better than the traditional telecom positioning approaches, it cannot compete with GPS. The main purpose of this paper is to present the methodology that can be applied to the estimated GPS locations by applying RF to achieve more accurate estimated locations.

2.1   Positioning Errors

With help of road network maps, we are able to observe the predicted points on the maps and two types of positioning errors as follows:

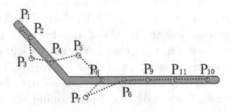

Fig. 2. Noise error points in a trajectory

As the Fig. 2 shows a trajectory consists of eleven points with noise.

– Horizontal error
  Horizontal error is not a simple error in latitude direction, and it mainly represents the predicted locations originally close to the road but now are far away from the true locations. In Fig. 2, there are three such examples: p_3, p_5, p_7. What we need to do is pull these errant points back to the road network.
– Vertical error
  Vertical error mainly represents the predicted location points in wrong sequences, although they are distributed on the correct road network. From the common sense, a human walking/driving trajectory will usually not appear in repeated crossing. The Fig. 2 depicts there are eleven points in the whole walking trajectory, and it should be expressed in the sequential order: $p_1 \rightarrow p_{11}$ according to the experience knowledge, but point p_8 appears between p_5 and p_7, and so point p_{11} does the same. We define this errant sequence as the vertical error.

To solve the above vertical and horizontal errors, we are going to leverage the road network maps and multiple consecutive GPS points as the context information to improve the positioning precision.

3   Telco Localization Solution Introduce

In this section, we first give an overview of several solutions that can correct errors in predicted locations [3].

3.1   Kalman Filter

Review of Kalman Filter. Before using Kalman Filter (KF) [1] as a tool to improve the positioning accuracy in our problem, we would like to first give a quick review of KF. More specifically, KF [7] mainly consists of two main parts: one is the state Eq. 1 and the other is the observation Eq. 2. The KF model assumes the true state at time k is evolved from the state at $(k-1)$ as Eq. (1) states.

$$\widehat{x}_k = Ax_{k-1} + Bu_k + w \tag{1}$$

where

- A is the system state parameter, which is the transition model applied to the previous state x_{k-1};
- B is the control-input model that is applied to the control vector u_k and can be ignored in this paper;
- w is the processing noise that is assumed to be zero mean Gaussian white noise, with covariance Q;
- At time k an observation (or measurement) z_k of the true state x_k is obtained according to the following observation equation.

$$z_k = Hx_k + v \tag{2}$$

where

- z_k is the observation result;
- H is the observation matrix;
- x_k is the true state value in its system;
- v is the observation noise that is assumed to be zero mean Gaussian white noise with covariance R (In this paper, we assumed that this covariance as well as Q won't be altered with the system state dynamically).

The updating equations from time $k-1$ to k are as follow.

$$\widehat{X}_k = AX_{k-1} + BU_k \tag{3}$$
$$\widehat{P}_k = AP_{k-1}A^T + Q \tag{4}$$
$$K_k = \widehat{P}_k H^T (H\widehat{P}_k H^T + R)^{-1} \tag{5}$$
$$\widehat{X}_k = \widehat{X}_k + K_k(Z_k - H\widehat{X}_k) \tag{6}$$
$$P_k = (1 - K_k H)P_k \tag{7}$$

where

- K_k is the Kalman gain at time k;
- P_k is the error covariance at time k.

Algorithm 1. Kalman Filter Algorithm

Input: Input GPS points ⟨ longitude, latitude, timestamp ⟩
Output: Corrected GPS points ⟨ longitude, lattitude, timestamp ⟩

1: set initial point and speed;
2: set noise Q and observation noise R;
3: **for** each input GPS point **do**
4:    compute the predicted position by Eq. 1;
5:    update the current position by Eq. 2;
6: **end for**
7: return points = sum($point_k$) k = 1,2,3.....;

Kalman Filter-Based Correction Algorithm. Algorithm 1 shows the overall Kalman filter procedure referred to two core equations as introduced before (1 and 2) to process the obtained training data from telco big data platform. First, we need to set a necessary initial data point and its speed according Kalman equations (line 1) as well as two noise sets Q and R noise (line 2). Second, for each sequential point (line 3) of the whole trajectory we should apply Kalman Filter to evaluate their true GPS values (lines 4 and 5, where we apply the series of equations (3 to 7) to compute the real time estimated values and update two kinds of noise mentioned above). Finally, the algorithm aggregates all evaluated points to form the original sequence (line 6). Equations 3 and 4 are to project the state and error covariance ahead, then compute kalman gain in Eqs. 5 and 6 and 7 update the estimation with measurements and error covariance.

3.2   Mean Filter-Based Correction

Recall that the KF-based correction algorithm does not fully leverage context information. In order to resolve the issue where the existing model-based algorithms are unable effectively to deal with the abnormal of predicted positions and increase the positioning accuracy, we borrow the idea of the mean filter [5] that has been applied in the imagery data processing and design a context-aware correction algorithm based on the GPS points inside a sliding window. The methodology will be applied to the post-processing of the predicted trajectory. Specifically, for a measured point position x_i, the estimate value of this point is the mean of its $n/2$ successive GPS points and $n/2$ proceeding GPS points, where n is the size of a given sliding window.

$$\hat{x}_i = \frac{1}{n} \left(\sum_{k=i-n/2}^{i+n/2} x_k \right) \qquad (8)$$

In the above equation, $\hat{x}_i$ is the estimate of x_i. To ensure that the mean filter-based correction algorithm work, we first need to preprocess the input GPS points with the equal interval interpolation. For the given input GPS points, we first find the minimum time interval between any two continuous points. Next, based on the minimal time interval, we will obtain these consecutive points with the time interval greater the minimal one, and fill the missed GPS points by the median interpolation.

In addition, we note that the mean filter is sensitive to the outliers contained in the input GPS points. To resolve this issue, we would like to find those outliers and remove them from the input GPS points. To find such outliers, we use a classical median filter.

Algorithm 2. Mean Filter-based Correction
_____
Input: Input GPS points $\langle$ longitude, latitude, timestamp $\rangle$, and window size n
Output: Corrected GPS points $\langle$ longitude, lattitude, timestamp $\rangle$
1: load the S input GPS points;
2: preprocess the GPS points by equal interval interpolation with help of a minimum time interval.
   eliminate outliers by using median filter;
3: **for** each i in size of window **do**
4:    **for** each of the n input GPS point **do**
5:       moving equalize point with the Eq. 3;
6:    **end for**
7: **end for**
8: points = $\text{sum}(point_k)$ k = 1,2,3.....;
_____

The body of this Algorithm 2 mainly depends on the Eq. 8 we introduced at the beginning of this section.

3.3 Map Matching-Based Correction

Map-matching [2,8] is to match the recorded geographic coordinates (such as collected GPS points) to a logical model of the real world. It has been developed as a very mature technique combining digital map with locating information, for example, to obtain the real position of vehicles in a road network.

We plug the map-matching technology to Algorithm 2 to filter out those GPS outliers during the preprocessing and postprocessing phases. First, in the preprocessing step in line 2 of Algorithm 2 it can plug in the map matching technique to make sure every input GPS point is on the correct road. In this way, we can remove the outliers in the input points. Second, even after Algorithm 2 is performed, it is still possible that some corrected GPS points (i.e., the output of Algorithm 2) might not be on the roads. Thus, we can again apply the map-matching technique to these corresponding GPS points to acquire the final corrected points or locations. In this case, Algorithm 2 can work together with the map-matching technique to improve the positioning accuracy significantly.

4 Evaluation

In this section, we compare the performances of three models: KF, mean-filter (MF) and map-matching (MF+MM) via the computational experiments. In order to measure the positioning precision, we first apply the model-based (random tree) positioning approach to derive the recovered or estimated GPS points from MR data. After that, we employ these models to correct the recovered GPS points. Based on the corrected GPS points by these three approaches, we then measure the positioning precision of each model. To perform the computational experiments, we use a real dataset of user mobility trajectories in one day (containing around 600,000 MR records) collected from a telecom service provider in the city of Shanghai, China.

4.1   Performance Comparison

In addition to the measurements of the positioning precisions of individual models mentioned above, we also include the positioning precision for the recovered GPS points as the baseline. We use two metrics, namely, the mean error and median error as shown in Table 1 to present the positioning precision.

Table 1. Comparison of recovered and corrected GPS points

Metric	Median (meters)	Mean (meters)
Recovered points	31.8075	56.6148
Corrected points by KF	31.2072	48.7996
Corrected points by MF	25.4192	39.3617
Corrected points by MF+MM	23.1236	31.2681

In Table 1, KF model slightly improves the recovered points' positioning precision. For example, the mean error is reduced from 56.6148 m to 48.7996 m and yet the median stays almost the same. From the table it can be seen that the median has a very little reduction from 31.8075 m to 31.2072 m. Albeit MF improves the positioning precision, MF+MM greatly increases the positioning precision with around 20.08% and 30.47% reductions in the mean and median errors respectively. The numbers in Table 1 clearly verify the superiority of MF+MM in terms of providing the positioning precision.

Next, we plot the error distribution of the recovered and corrected GPS points in Fig. 3. The x-axis represents the error range and the y-axis represents the error distributions in the different error ranges. Figure 3(a) shows the original error distribution without any error correction process. Figure 3(b) depicts the error distribution of each recorded point that is simply corrected by the KF model. Nevertheless, the resultant error distribution is very similar to the original one. The prerequisite of classic Kalman filtering is to establish an accurate dynamic model and observation model, and it needs more clearly understanding of the moving object [4]. But in this paper, we assume that the observation equation is linear and with the stable noise.

Figure 3(c) represents the error distribution for the results obtained by MF mode. This algorithm has a relatively good effect for the wide range of error

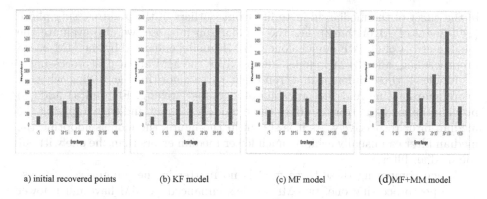

a) initial recovered points   (b) KF model   (c) MF model   (d)MF+MM model

Fig. 3. Whole error distribution

correction because it is more inclined to smoothing trajectory based on the empirical knowledge of human motion behavior and maintains a sustained and stable state for a short time of period. It can be seen from Fig. 3(c) the number of points with the biggest errors has reduced dramatically compared to the original one. Similarly, this approach also helps correcting the errors for these locations around the point that has relatively big error, and these locations usually can be easily affected by this outlier. Figure 3(d) reflects the combined effect about MF and Map matching. From the picture we are able to recognize that this approach is more effective in reducing errors after introducing map-matching method.

4.2 Sensitive study of KF+MM model

We can use the map-matching in both pre-processing and post-processing phases. To study the sensitivity of the KF+MM model, we vary the window size and measure the median error of the KF+MM model. In addition, we are interested

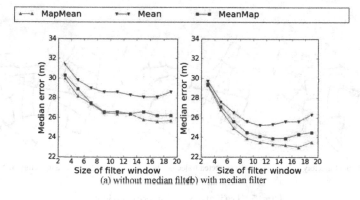

(a) without median filter   (b) with median filter

Fig. 4. Error curve changing window size

in the effect of the median filter to filter out outliers in the preprocessing phase (i.e., line 2 of Algorithm 2). For the comparison purpose, we also evaluate the median errors of (1) MF model alone, (2) MF model enhanced by the MM-based preprocessing, and (3) MF model enhanced by the MM-based postprocessing.

In Fig. 4, three different lines represent three algorithms according whether adopt map-mating and the order adopted map-matching. Figure 4(a) and (b) provide the median errors of above three models without and with applying the median filter respectively. As it is shown in the figure, the models adopting the median filter can usually achieve much lower median errors than the one without the median filter.

When comparing these three models no matter in the preprocessing phase or the postprocessing one, two MF models enhanced by MM have much lower median errors than the MF model alone. Moreover, we find that the MF model with the MM-based preprocessing out performs the one with the MM-based postprocessing. It is because the MM-based preprocessing can clean the outliers without affecting the MF model. In MF model with MM-based postprocessing, the outliers still appear in the input of the MF model and could impact the overall precision of the KM model negatively.

4.3   Visualization of Positioning Models

Figure 5 illustrates the recovered GPS points and corrected points or locations on a real road network. First in Fig. 5(a), there exist many points that are not on roads. It is mainly because the two-layer random forests (RFs) use the center points of those leaf nodes in the RFs as the predicted GPS points, no matter the center points are on the roads or not. Second, Fig. 5(b) demonstrates that MF model obviously is able to smooth the trajectory. Nevertheless, in our dataset, many GPS points appear on the overpass and underpass that occur on the same street segments. It is hard for the MF model to put them onto the correct roads (because the roads on the overpass and underpass share the same longitude and latitude coordinates). However, Fig. 5(c) demonstrate that this problem can be

(a) original trajectory   (b) corrected trajectory by MF   (c) corrected trajectory by MF+MM

Fig. 5. Trajectory error comparison

overcome by applying the MM technique and as the result it produces the best positioning precision.

5 Conclusion

In this paper, we propose the methodology to leverage context information in order to correct the estimated or recovered city-scale localization errors by applying model-based localization methods. By adopting the powerful techniques including Kalman Filter, Mean filter, and Map-matching, the proposed approaches can greatly improve the positioning precision.

As the future work, we will continue work on improving the positioning precision. For example, we are planning to explore the regularity patterns from the recovered trajectories for further postprocessing. In addition, beyond the regression model-based prediction algorithm, we are interested in other advanced machine learning techniques to replace the currently used random forest models.

Acknowledgements. The authors would like to appreciate Professor Weixiong Rao for his critical and useful suggestions. The authors also want to thank two anonymous reviewers for their critics and suggestions that help improving the quality of our paper.

References

1. https://en.wikipedia.org/wiki/kalman_filter
2. https://en.wikipedia.org/wiki/map_matching
3. Anderson, J.L.: Exploring the need for localization in ensemble data assimilation using a hierarchical ensemble filter. Phys. D: Nonlinear Phenom. **230**(1), 99–111 (2007). Data Assimilation
4. Anderson, J.L.: Localization and sampling error correction in ensemble kalman filter data assimilation. Mon. Weather Rev. **140**(7), 2359–2371 (2012)
5. Dawoud, N.N., Samir, B.B., Janier, J.: N-mean kernel filter and normalized correlation for face localization, pp. 416–419 (2011)
6. Dil, B., Dulman, S., Havinga, P.: Range-based localization in mobile sensor networks. In: Römer, K., Karl, H., Mattern, F. (eds.) EWSN 2006. LNCS, vol. 3868, pp. 164–179. Springer, Heidelberg (2006). doi:10.1007/11669463_14
7. Evensen, G.: The ensemble kalman filter: theoretical formulation and practical implementation. Ocean Dyn. **53**(4), 343–367 (2003)
8. Lou, Y., Zhang, C., Zheng, Y., Xie, X., Wang, W., Huang, Y.: Map-matching for low-sampling-rate GPS trajectories, pp. 352–361 (2009)
9. Yuan, M., Deng, K., Zeng, J., Li, Y., Ni, B., He, X., Wang, F., Dai, W., Yang, Q.: OceanST: a distributed analytic system for large-scale spatiotemporal mobile broadband data. PVLDB **7**(13), 1561–1564 (2014)
10. Zhu, F., Luo, C., Yuan, M., Zhu, Y., Zhang, Z., Gu, T., Deng, K., Rao, W., Zeng, J.: City-scale localization with telco big data. In: Proceedings of the 25th ACM Conference on Information and Knowledge Management, CIKM 2016, 24–28 October 2016, Indianpolis, USA (2016)

Inferring Unmet Human Mobility Demand with Multi-source Urban Data

Kai Zhao[1], Xinshi Zheng[1], and Huy Vo[2(✉)]

[1] Center for Urban Science and Progress, New York University, New York City, USA
{kai.zhao,xinshi.zheng}@nyu.edu
[2] Department of Computer Science, City College of the City University of New York,
New York City, USA
huy.vo@nyu.edu

Abstract. As the sharing economy has been increasing dramatically in the world, the mobile-hailed ridesharing companies like Uber and Lyft in the US, Didi Chuxing in China has begun to challenge traditional public transportation providers such as bus, subway or taxis. Ridesharing companies have shown their ability to provide the mobility services where public transit has failed. The human mobility demand that cannot be satisfied by traditional transportation modes (unmet human mobility demand) can be served by the ridesharing companies. In this paper, we provide a 'hydrological' perspective for inferring unmet mobility demand patterns in cities with multi-source urban data. We observe that the unmet human mobility demand is proportional to the met mobility demand by examining the yellow taxi and the Uber data in New York City. Based on this observation, a Single Linear Reservoir (SLR) model has been proposed for modeling unmet human mobility demand from multi-source urban data.

Keywords: Urban human mobility · Spatio-temporal data mining

1 Introduction

As the sharing economy has been increasing dramatically in the world, the mobile-hailed ridesharing companies like Uber and Lyft in the US, Didi Chuxing in China has begun to challenge traditional public transportation providers such as bus, subway or taxis. Ridesharing companies have shown their ability to provide the mobility services where public transit has failed. The human mobility demand that cannot be satisfied by traditional transportation modes (unmet human mobility demand) can be served by the ridesharing companies [10].

In this paper, we provide a 'hydrological' perspective for inferring unmet mobility demand patterns (e.g., Uber) in cities with multi-source urban data (e.g., subway, taxi, bus, or bike data). We observe that the unmet human mobility demand is proportional to the met mobility demand by examining the yellow taxi and the Uber data. A SLR model [3] developed previously for rainfall-runoff analysis has been adopted for modeling unmet mobility demand. The primary

S. Song et al. (Eds.): APWeb-WAIM 2017 Workshops, LNCS 10612, pp. 118–127, 2017.
https://doi.org/10.1007/978-3-319-69781-9_12

goal of this paper is to establish the mathematical relationship between the urban human mobility demand and the unmet mobility demand, and hence develop a 'unit hydrograph' methodology for predicting future unmet mobility demand.

In this paper, urban human mobility status has been simplified with the following assumptions: (1) The mobility demand can be met by any type of public transportation including taxi, bus, subway and bike; (2) The demand cannot be met for a given time period will eventually be met by ridesharing transportation modes such as Uber. The total urban human mobility demand is defined as the number of passengers in a region that have travelling demand using ground transportation, such as subway, taxi, bike or bus, during a given time interval. The met urban human mobility demand is the amount of passengers that are able to find a method of transportation within this given time period. The unmet mobility demand is the number of passengers that cannot find any transportation mode during this given time period.

First, we observe that the unmet human mobility demand is proportional to the met mobility demand. Based on this observation, we borrow the 'unit hydrograph' concept in hydrology, which is originally the unit pulse response of runoff for a watershed receiving a unit amount of excess rainfall for a given duration, is defined here as the response of the number of passengers to choose Uber given a unit input of urban human mobility. However, unlike hydrological studies, which are focused on deriving the 'unit hydrograph' to estimate the runoff from rainfall input, our aim is to use this method to estimate the change of storage within the linear reservoir when the total urban human mobility demand changes. The reason we use the SLR model is that, the urban human mobility demand [10] is similar to the rainfall-runoff [3]. It is dynamic with spatially distributed inputs and outputs. There is a peak of the human mobility demand, like the rainfall input, and a concentration time that the human mobility demand to be full-filled, like the time the water needs to pass through the system. The unmet human mobility demand, similar to the runoff water, is the amount of passengers plan to travel from one location to another, but cannot find the right public transportation system.

2 Positive Correlation Between the Taxi and Uber Demand

Ridesharing companies have shown their ability to provide the mobility services where public transit such as taxi or bus has failed. The human mobility demand that cannot be satisfied by traditional transportation modes, i.e., the unmet human mobility demand, can be served by the ridesharing companies such as Uber or Lyft. In this section, we show that the unmet human mobility demand is proportional to the met mobility demand by examining the yellow taxi and the Uber data in New York City [12]. The yellow taxi and Uber utilize different cruising strategies. The yellow taxis usually use a random cruising strategy, while the Ubers can go to the passenger's places when a request is received. Therefore, when traditional transportation modes fail to meet the human mobility demand, ridesharing companies such as Uber have the potential to satisfy the unmet human mobility demand.

2.1 Data-Sets

First, we give a description of the data-sets used in this paper and how we pre-process the data:

Taxi and Uber Data-Sets. The NYC yellow taxi data-set is a public data-set provided by the Taxi and Limousine Commission (TLC) [7]. TLC records the information from all trips completed in yellow taxis in NYC. Each trip record includes fields capturing pick-up and drop-off time, pick-up and drop-off locations, trip distances, itemized fares, and passenger counts. In total we have 13,813,031 taxi pick-up records from 13,237 yellow taxis.

The New York Uber data-set is a public data-set from TLC [7] that aims to study the Uber behaviour. This data-set contains 663,845 Uber pick-up records. We examine both the yellow taxi and Uber data-sets for one month (June 2014). We extract following information from the data-set: taxi ID, pick-up time and the corresponding pick-up location (neighborhood).

Data Pre-processing. We use the NYC neighborhood (in total 184 neighborhoods) shape file [6] to map the pick-up GPS points with the associated neighborhoods: if the pick-up location is within the neighborhood, we consider that neighborhood as the one passengers getting on the taxi and there is one mobility demand at that neighborhood. All of our data pre-processing were conducted using the operational data facility at our research center [4,5]. In particular, the mapping of taxi pick-ups to geospatial features, which requires a lot of processing given the volume of the pick-up trips, on a 1200+ core cluster running Cloudera Data Hub 5.4 with Apache Spark 1.6. The cluster consists of 20 high-end nodes, each with 24 TB of disk, 256 GB of RAM, and 64 AMD cores.

2.2 Linear Correlation Between Met and Unmet Human Mobility Demand

We find that there is a strong positive correlation between the yellow taxi (met mobility demand) and Uber (unmet mobility demand) pick-ups (see Fig. 1(a)). Figure 1(b) shows the density of hourly pick-ups of yellow and Uber in the 184 neighborhood. A high density of points are near diagonal line, identifying a clear positive correlation.

We use the Pearson correlation coefficient to quantify the strength of the correlation between the hourly pick-ups of yellow and Ubers in all 184 neighborhoods. Our observation verifies our proof in the SLR model with a Pearson values as 0.82. The p−value is less than 0.01, identifying a very strong statistical significance. We show that there is a strong positive correlation between the unmet human mobility demand served by Uber and the met mobility demand served by yellow taxi in New York City.

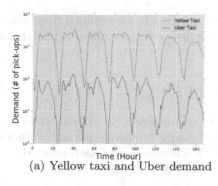
(a) Yellow taxi and Uber demand

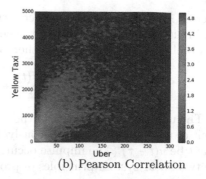
(b) Pearson Correlation

Fig. 1. (a) One week met mobility demand (yellow taxi) and unmet mobility demand (Uber) in June, 2014. (b) Positive correlation between met mobility demand and unmet mobility demand. (Color figure online)

3 Linear Reservoir Model for Unmet Mobility Demand Estimation

Since we observe that the unmet human mobility demand is proportional to the met human mobility demand by examining the yellow taxi and the Uber data. In this section, we propose a Single Linear Reservoir (SLR) model for modeling unmet mobility demand based on this observation. The primary goal is to establish the mathematical relationship between the urban human mobility demand and the unmet mobility demand, and hence develop a 'unit hydrograph' methodology for predicting future unmet mobility demand.

The reason we use the SLR model is that, urban human mobility demand phenomena is similar to the rainfall-runoff pattern in hydrology. It is dynamic with spatially distributed inputs and outputs. The system is regarded as a single idealized reservoir with unmet mobility demand as storage (S). The human mobility demand D (A in discrete time scale) is the input (rainfall) into the reservoir, and the portion of mobility demand that will be met within a specified time interval will be the output (runoff) of the reservoir (M). Here the human mobility demand A can be inferred with multi-view learning algorithm [9].

The linear reservoir model will first be analyzed in continuous time scale, and then converted to discrete time scale for application in developing 'unit hydrograph' for unmet mobility demand analysis. For an ideal linear system, there is continuity equation on continuous time scale:

$$\frac{dS(t)}{dt} = D(t) - M(t) \tag{1}$$

where t is time.

Since the system is linear, it is reasonable to assume that the unmet mobility demand is proportional to the output, the met mobility demand. There is:

$$S(t) = kM(t) \tag{2}$$

where k is a constant response factor that can either be determined from historical input, output data, or from the characteristics of the studied urban region.

Combining Eqs. (1) and (2), there is:

$$k\frac{dM(t)}{dt} + M(t) = D(t) \tag{3}$$

The unit impulse response of the system occurs when the system receives an input of unit amount instantaneously. Let the unit impulse response function at time t be $u(t - \tau)$ (The impulse occurs at τ). There is convolution integral from the two linear system principles of proportionality and superposition:

$$M(t) = \int_0^t D(\tau)u(t - \tau)d\tau \tag{4}$$

The unit step response of a system $r(t)$ is resulted from an input that changes from 0 to 1 at time 0 and continues indefinitely at that rate thereafter. With Eq. (4) $r(t)$ is found for $D(\tau) = 1$ for $\tau \geq 0$:

$$r(t) = \int_0^t u(t - \tau)d\tau = \int_0^t u(l)dl \tag{5}$$

where $l = t - \tau$.

The unit pulse response function $p(t)$, which is resulted from an input of unit amount occurring in duration Δt, can be determined based on the two linear system principles:

$$p(t) = \frac{1}{\Delta t}[r(t) - r(t - \Delta t)] = \frac{1}{\Delta t}\int_{t-\Delta t}^t u(l)dl \tag{6}$$

Similar to hydrological data, urban mobility data will be in discrete time intervals. In discrete time intervals of duration Δt, for the input of the system there is:

$$A_i = \int_{(i-1)\Delta t}^{i\Delta t} A(\tau)dt \tag{7}$$

where A_i is the accumulated urban human mobility demand during the time interval Δt. And $i = 1, 2, 3, ..., I$, where I is the last time interval of Δt. $D(\tau) = D_i/\Delta t$ for $(i - 1)\Delta t \leq \tau \leq i\Delta t$. And $D(\tau) = 0$ for $\tau > I\Delta t$.

The model output will be recorded differently, using the met mobility demand at the end of jth time interval M_j as the output for the jth time interval:

$$M_j = M(j\Delta t) \tag{8}$$

The unit pulse response at $t = j\Delta t$ from an input of duration Δt ending at $(i-1)\Delta t$ is found by Eq. (6):

$$p[t - (i-1)\Delta t] = p[(j - i + 1)\Delta t] = \frac{1}{\Delta t} \int\limits_{(j-i)\Delta t}^{(j-i+1)\Delta t} u(l)dl \qquad (9)$$

For $t \geq I\Delta t$, convolution integral Eq. (4) can be broken down into I parts:

$$M_j = \int\limits_0^{j\Delta t} D(\tau)u(j\Delta t - \tau)d\tau = \sum_{i=1}^{I} \frac{A_i}{\Delta t} \int\limits_{(i-1)\Delta t}^{i\Delta t} u(j\Delta t - \tau)d\tau \qquad (10)$$

In each of the I integrals, there is $l = t - \tau = j\Delta t - \tau$, together with Eq. (9), for the jth integral there is:

$$\frac{A_i}{\Delta t} \int\limits_{(i-1)\Delta t}^{i\Delta t} u(j\Delta t - \tau)d\tau = \frac{A_i}{\Delta t} \int\limits_{(j-i+1)\Delta t}^{(j-i)\Delta t} -u(l)dl$$

$$= \frac{A_i}{\Delta t} \int\limits_{(j-i)\Delta t}^{(j-i+1)\Delta t} u(l)dl = A_i p[(j - i + 1)\Delta t]' \qquad (11)$$

Let $U_{j-i+1} = p[(j - i + 1)\Delta t]$, Eq. (11) can become:

$$M_j = \sum_{i=1}^{I} A_i U_{j-i+1} \qquad (12)$$

Similarly, for $t < I\Delta t$, the output can be divided into j parts at time $t = j\Delta t$ written as:

$$M_j = \sum_{i=1}^{j} A_i U_{j-i+1} \qquad (13)$$

Combining Eqs. (12) and (13), there is:

$$M_j = \sum_{i=1}^{min[I,j]} A_i U_{j-i+1} \qquad (14)$$

Equation (14) can be further expressed in matrix form:

$$[A][U] = [M] \qquad (15)$$

Linear regression can be used to derive U for Eq. (15) given A and M. Assume an estimate will be found for U that yields $[\hat{M}]$. The solution can be found with least square error minimization between $[M]$ and $[\hat{M}]$. The solution will be:

$$[U] = [[A]^T[A]]^{-1}[A]^T[M] \tag{16}$$

Equation (1) can be rewritten in discrete time as:

$$S_j - S_{j-1} = A_j - M_j \tag{17}$$

where S_j is the unmet mobility demand at the end of the jth time interval. And $A_j = 0$ for $j > I$.

Assume the initial unmet mobility demand is zero ($S_0 = 0$), iteratively, Eq. (17) can become:

$$S_j = \sum_{j=1}^{min[I,j]} A_j - \sum_{j=1}^{j} M_j \tag{18}$$

4   Discussion

To use this SLR model to estimate the unmet mobility demand, three types of data will be needed for a given region: (i). Historical data of total human mobility demand from multiple sources as highlighted (A_{hist}); (ii). Historical data of met mobility demand (M_{hist}); (iii). Future prediction data of total human mobility demand (A_{future}). Use A_{hist} and M_{hist} the 'unit hydrograph' of mobility demand (U) in this region can be found using least-squares fitting with Eq. (16). The obtained U will be used with A_{hist} for Eq. (14) to estimate the future met mobility demand M_{future}. Finally, with Eq. (18) the future unmet mobility demand S_{future} can be estimated.

We provide a recommended work-flow to use the 'unit hydrograph' approach in unmet mobility demand prediction (also see Algorithm 1):

1. Delineate the studied city into sub-areas, so that within each sub-area the total travelling demand is roughly uniformly distributed;
2. For each of the sub-areas, collect multi-source data of total mobility demand and demand that has been met based on relevant historical data-source and/or results from other models;
3. Within each sub-region, develop the 'unit hydrograph';
4. Use the 'unit hydrograh' with the predicted/estimated future mobility demand data to compute the future mobility demand that will be met;
5. Use the computed future mobility demand that will be met together with the future total mobility demand, the future unmet mobility demand can be estimated.

Since this is still an ongoing work, we did not implement and compare the SLR model with other unmet demand estimation algorithms [1,2]. In future work, we will conduct an experiment in NYC, using multi-source data-sets to

Algorithm 1. Linear Reservoir Model for Unmet Mobility Demand Estimation

| input | : The historical urban human mobility demand A_i with each time intervals Δt for a particular region, the met urban human mobility demand M_j with the same time and spatial frame as A_i from multi-source historical data, including taxi pick-ups, public transportation usage, bike usage, ect. |

output: The future unmet mobility demand S_j

1 Use deconvolution and linear regression to develop the 'unit hydrograph' U with historical A and M data;
2 Use derived U from step 1 to predict future M_j by equation (15), given that future A known (Future A can be estimated from arbitrary transportation demand models);
3 Use A_i and M_j to compute the predicted unmet mobility demand S_j iteratively for each time interval.

validate the accuracy of the model as well as calibrate it (see Fig. 2). We believe such a model will provide us more insights in understanding the urban human mobility demand problem, and provide decision support for the design of urban transportation system.

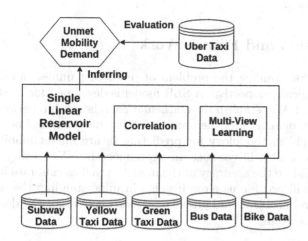

Fig. 2. Framework for inferring unmet mobility demand.

5 Related Work

5.1 Mobility Modeling Based on Multi-source Urban Data

Existing urban human mobility are mostly driven by data from a single view, e.g., data from a single transportation view such as taxi or subway. The study based on the single-source data inevitably introduces a bias against city residents not contributing this type of data, e.g., residents who walk [8] or ride private vehicles. To address this issue, Zhang et al. [9] propose a human mobility model based on multi-source urban data. They introduce a multi-view learning framework and observe that the model outperforms a single-view model by 51% on average. Zhao et al. [11] decompose the human mobility trips into different classes according to different transportation modes, such as Walk/Run, Bike, Train/Subway or Car/Taxi/Bus. They observe that human mobility can be modelled as a mixture of different transportation modes, and these single transportation movement patterns can be approximated by a log-normal distribution.

5.2 Inferring Unmet Taxi Demand

Recent papers try to infer the unmet taxi demand, the number of people who need a taxi but could not find one, from the taxi data-set. In [2] the authors combine flight arrival with taxi demand and predict the passenger demand at different airport terminals in Singapore use queuing theory. Anwar et al. [1] formalize the unmet taxi demand problem and present a novel heuristic algorithm to estimate it without any additional information. They infer the unmet taxi demand from taxis with empty services and show that it can be used to quantify the unmet demand.

6 Conclusion and Future Work

In this paper we examine the problem of predicting unmet mobility demand with a hydrological perspective. A SLR model is developed for modeling unmet mobility demand. We establish the mathematical relationship between the urban human mobility demand and the unmet mobility demand, and hence develop a 'unit hydrograph' methodology for predicting future unmet mobility demand. In the next step, we will conduct an experiment in NYC, using multi-source data-sets to validate the accuracy of the model as well as calibrate it. We believe such a model will provide us more insights in understanding the urban human mobility demand problem, and provide decision support for the design of urban transportation system.

Acknowledgment. The authors thank: the New York City TLC for providing the data used in this paper. This work was supported in part by a CUNY IRG Award and the NYU Center for Urban Science and Progress.

References

1. Afian, A., Odoni, A., Rus, D.: Inferring unmet demand from taxi probe data. In: 2015 IEEE 18th International Conference on Intelligent Transportation Systems, pp. 861–868, September 2015
2. Anwar, A., Volkov, M., Rus, D.: Changinow: a mobile application for efficient taxi allocation at airports. In: 2013 IEEE 16th International Conference on Intelligent Transportation Systems, pp. 694–701, October 2013
3. Chow, V.T., Maidment, D.R., Mays, L.W.: Applied Hydrology (1988)

4. Freire, J., Bessa, A., Chirigati, F., Vo, H.T., Zhao, K.: Exploring what not to clean in urban data: a study using new york city taxi trips. IEEE Data Eng. Bull. **39**(2), 63–77 (2016)
5. Miranda, F., Doraiswamy, H., Lage, M., Zhao, K., Gonçalves, B., Wilson, L., Hsieh, M., Silva, C.T.: Urban pulse: capturing the rhythm of cities. IEEE Trans. Vis. Comput. Graph. **23**(1), 791–800 (2017)
6. New York Open Data Set. http://www1.nyc.gov/site/planning/data-maps/open-data.page
7. New York Taxi Data Set. http://www.nyc.gov/html/tlc
8. Rao, W., Zhao, K., Zhans, Y., Hui, P., Tarkoma, S.: Maximizing timely content advertising in DTNs. In: 9th Annual IEEE Communications Society Conference on Sensor, Mesh and Ad Hoc Communications and Networks, SECON 2012, Seoul, Korea (South), 18–21 June 2012, pp. 254–262 (2012)
9. Zhang, D., Zhao, J., Zhang, F., He, T.: comobile: real-time human mobility modeling at urban scale using multi-view learning. In: SIGSPATIAL, Bellevue, WA, USA, 3–6 November, pp. 40:1–40:10 (2015)
10. Zhao, K., Khryashchev, D., Freire, J., Silva, C.T., Vo, H.T.: Predicting taxi demand at high spatial resolution: approaching the limit of predictability. In: 2016 IEEE International Conference on Big Data, BigData 2016, Washington DC, USA, 5–8 December 2016, pp. 833–842 (2016)
11. Zhao, K., Musolesi, M., Hui, P., Rao, W., Tarkoma, S.: Explaining the power-law distribution of human mobility through transportation modality decomposition. Nat. Sci. Rep. **5**(9136) (2015)
12. Zhao, K., Tarkoma, S., Liu, S., Vo, H.T.: Urban human mobility data mining: an overview. In: 2016 IEEE International Conference on Big Data, BigData 2016, Washington DC, USA, 5–8 December 2016, pp. 1911–1920 (2016)

A Hybrid Approach of HTTP Anomaly Detection

Yang Shi, Shupei Wang, Qinpei Zhao[✉], and Jiangfeng Li[✉]

School of Software Engineering, Tongji University, Shanghai, China
{qinpeizhao,lijf}@tongji.edu.cn

Abstract. Security technology in computer network including anomaly detection is increasingly playing an important role in the government and protection of Internet along with its popularity. Anomaly detection uses data mining techniques to detect the unknown malicious behavior. Various hybrid approaches have been proposed in order to detect outliers more accurately recently. This paper proposes a novel hybrid of clusterings and graph to detect anomaly. We introduce a new holistic approach in a common bipartite scenario of users from intranet accessing to Internet that utilizes different types of clusterings for the individual feature data to find the outliers and then a graph model to take advantage of the relational data naming network to enhance anomaly detection. The framework solution has several advantages: taking consideration of individual feature data and relational data, keeping open to extend different types of clusterings, easily appending more domain knowledge.

Keywords: Anomaly detection · Data mining · Clustering · Bipartite graph

1 Introduction

Network traffic anomalies are unusual and significant changes in terms of network traffic amount. Given a large amount of network traffic, how to precisely detect anomalies is very challenging. In particular, it is hard to get an enough amount of ground truth labels to identify whether one is normal or abnormal. Classifiers usually work with labels. Instead clusterings do not need any label data, and have been used for network anomaly detection. Indeed, the classification algorithms often can get a higher detection rate than clusterings.

To overcome the issue above, in this work, we are inspired by [9], and propose to integrate the graph based anomaly inference and cluster ensemble techniques to take full advantage of feature data and network behaviour relational data.

Specially, we first adopt clustering ensemble to combine the strength of individual clustering algorithms (such as K-means, DBSCAN, Birch and Meansshift). With the clustering ensemble of such four clustering algorithms, we can label a small amount of anomalies with high confidence. Next, we model the relations between HTTP clients and servers as a bipartite graph, and apply Markov

© Springer International Publishing AG 2017
S. Song et al. (Eds.): APWeb-WAIM 2017 Workshops, LNCS 10612, pp. 128–137, 2017.
https://doi.org/10.1007/978-3-319-69781-9_13

Random Field (MRF) on the bipartite graph to infer that the nodes are normal or abnormal. Finally, we solve the inference problem by Loopy Belief Propagation (LBP). Figure 1 illustrates the overall structure of hybrid of clustering and graph-based inference.

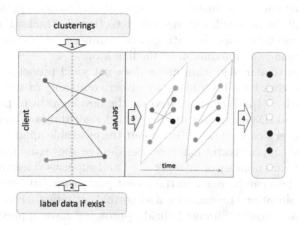

Fig. 1. Overall structure of hybrid of clusterings and graph.

Our main contributions are as follow: (1) We introduce a novel approach for network anomaly detection problem, which ties together relational data with high dimension individual information, it utilizes all of graph, (2) We recommend clustering as the predictor of prior potential, that could be applied as a general method in others unknown network and domain and keep open to extend different types of clusterings, (3) Our work is in a completely semi-supervised fashion and amendable in situation of better prior value or lack of prior.

We evaluate our method on real world data from a Internet security company. Verification on known data provided by the company in the semi-supervised fashion is more than 80% accuracy rate of malicious inference. The improvement of accuracy rate with respective to clustering algorithm is about 20 %.

2   Related Works

Different techniques in network security domain or data mining had taken part in the mission of network anomaly detection on their perspective. The signature-based approaches look for pattern that matches known signatures. For example, dos activities can be detected based on the uniformity of ip address [5]. While the traditional signature-based detection will face great challenges as they will likely be outpaced by the threats created by anomaly authors [7]. The non-signature based approach, naming statistical techniques are applied, too. These techniques analyse the individual information. Classification and cluster algorithm can detect the network anomalies taking account of collective information of more features. The previous works [2,4] use classification algorithm to detect network anomalies. [2] choose the One Class Neighbor Machine and the recursive Kernel-based Online Anomaly Detection algorithm to detect anomalous behaviour in a distributed set of network measurements.

Some previous literatures has used clustering ensemble for anomaly detection. In [6], they apply ensemble clustering to anomaly detection, hypothesizing that multiple views of the data will improve the detection of attacks. Each clustering rates how anomalous a point is; ratings are combined by averaging or taking either the minimum, the maximum, or median score.

Using any particular algorithm alone does not yield proper results. In past few years approaches have been made by either combining or merging different algorithms together [1]. [3] concentrated on the development of performance of naive bayesian and ID3 algorithm and his hybrid algorithm was tested in Knowledge Data Discovery cup. [8] described about the ensemble approach which used decision tree and support vector machine. Besides the above cascading supervised techniques, there are number of unsupervised and supervised learning algorithm whose combinations can be made in the recent past years. [10] combine k means and ID3 for classification of anomalous and normal activities in computer address resolution protocol traffic. Similar hybrid approaches have applied in the SVM classification and k medoids clustering, k medoids clustering and naive bayes classification, one class and two class SVM, etc.

The classification and clustering algorithms, however ignore the network relational information when detecting the anomalies. Graph mining methods, which take advantage of relational information, have been successfully applied in many domain [7] from authority propagation to fraud detection. In [9], a new holistic approach that utilizes clues from all meta data as well as relational data to spot suspicious users and reviews in a online review system was proposed. In [7], they formulate the classic malware detection problem as a large-scale graph mining and inference problem, and show how domain knowledge is readily incorporated into the algorithm to identify malware.

3   Methodology

3.1   Clustering

Clustering can be defined as a division of data into group of similar objects. its advantages is able to detect anomaly without prior knowledge. Beyond the robustness provided by each algorithm, *Clustering ensemble* has been recently considered in variety of different areas used as a meta-clustering method to improve the robustness of clustering by combining the output of multiple algorithm. It can be defined as the optimization problem where, given a set of m clusterings, we want to find the clustering that minimizes the total number of disagreements with the m clusterings. There are various methods of clustering that can be applied for the anomaly detection. Intuitively, different types of clusterings or different parameters of one clustering are picked up to attain anomaly from a diverse perspective. Each chosen clustering algorithm is regarded as "anomaly voter". Meanwhile each node is detected and classified as anomalous or normal by "anomaly voters". Some node is detected to be anomaly by more voters, it's assigned a higher suspicious score, while some node is detected to be anomaly by less voters, it's assigned a lower suspicious score. The rest nodes, which are not detected by any voters, are regarded as normal with a high unsuspicious score.

$$score(node_i) = \begin{cases} a, & if \quad voter = 0 \\ 0.5 + 0.5 * (voter/N). \end{cases} \quad (1)$$

The Eq. 1 essentially transfers the clusterings binary result to a numerical value (the parameter a is constant and less than 0.5, N is the total number of "anomaly voter", voter is the number of "anomaly voters" which regard the node as anomalous). The value is initial prior probability of each node's goodness or badness.

It's critical and skillful to balance the amount of anomaly detected by different candidate clusterings. We firstly take advantage of the high efficiency of BIRCH with fast running time, run a large number of experiments with different radius parameter, obtain the distribution of anomaly quantity and choose the proper anomaly quantity. Following the clustering result of Birch, we will tune the parameter for the rest of three clustering algorithms. For fairness, we ensure that the number of outliers by each clustering algorithm is approximately equal.

3.2 MRF Model

Review of MRF. Markov Random Field (MRF) is a class of graphical models particularly suited for solving inference problems with uncertainty in observed data. Markov network or undirected graphical model is a set of random variables having a Markov property described by an undirected graph. The Markov property in Markov random field is that the probability distribution of one node's state depends on the nodes' state surrounding it and not on other nodes. In a MRF, each node is in any of a finite number of states. The dependency between a node and its neighbors is represented by a *propagation matrix* (ψ), where $\psi(i, j)$ equals the probability of a node being in state j given that it has a neighbor in state i.

Problem Description. We first give the intuition of using MRF of anomaly detection. When attackers initiate network traffic attacks and cause anomaly on target nodes, the target nodes' neighbors (and also neighbors of neighbors and etc.) could be harmed by the attacks. Thus, the network traffic behaviour of such neighbors could deviate from the majority of normal nodes. With help of MRF, we can detect how network traffic is prorogated across node edges to detect network traffic anomalies by aggregating network traffic from neighbor nodes (together with neighbors of neighbors and etc.) towards target nodes. In a nutshell, we adopt MRF to the client-server bipartite graph in Fig. (1) and formulate the anomaly detection problem as a network classification task.

An MRF model consists of an undirected graph where every node i is associated with a random variable Y_i that can be in one of a finite number of states or class represented by labels. The label of nodes is assumed to be dependent only on these nodes surrounding with edge and independent on all the other nodes in the network. The joint probability of node is written as a product of individual and pairwise factors:

$$P(y) = \frac{1}{Z} \prod_{Y_i \in V} \varphi_i(y_i) \prod_{(Y_i, Y_j) \in E} \psi_{ij}(y_i, y_j) \tag{2}$$

where y refers to an assignment of labels of all nodes, y_i denotes node i's assigned label and Z is the normalization constant which make the probability of all states of one node sum up to 1. The individual factors φ is called prior potentials and represent initial class probabilities for every node, often initialized based on prior knowledge. For the prior potentials, one can estimate them by proprietary formula or using external information. Prior potentials are obtained mainly from the clustering score.

In our whole design and implementation, one of the goal is to require the credible labels through combination of general and common techniques rather than proprietary formula in some field. The pairwise factors ψ are called compatibility potentials and capture the likelihood of a node with label y_i to be connected to node with label y_j.

Table 1. Compatibility potential matrix

	Server	
Client	Abnormal	Normal
Abnormal	$0.5 + \varepsilon$	$0.5 - \varepsilon$
Normal	$0.5 - \varepsilon$	$0.5 + \varepsilon$

The compatibility potentials are set based on several intuitions reflecting the modus-operandi of abnormal nodes. The anomalous server usually access the anomalous clients with larger probability than with normal clients, likewise the normal servers connect to the normal clients with larger probability. The compatibility potential of client and server node is as follow in Table 1 (ε is set 0.01), it's elegant form of the compatibility potentials ψ and convenient for calculating Eq. (2).

Given the model parameter (φ_i and $\psi_{i,j}$), the task is to infer the maximum likelihood assignment of states to the random variables associated with the nodes, naming to find the y that maximizes the joint probability of the network. The inference problem is an NP-hard task. The enumeration of all possible states is exponential and thus intractable for large network.

Fortunately, the *belief propagation* algorithm has been proven very successful in solving inference problem over graph in various field (e.g. image restoration, error-correcting code). In particular, the iterative message passing in BP

intuitively simulate the client-server request and response behaviour and network traffic aggregation toward target nodes. Thus, iterative approximate inference algorithm such as *Loopy Belief Propagation* (LBP) is a good choice for our project. In the next section, we describe the details of LBP to solve our problem.

3.3 LBP Algorithm

LBP infers the label of nodes from prior potential of the node and compatibility potential involving with the node's neighbours through iterative message passing between all pairs of node i and j. $m_{i,}(x_j)$ denote the message sent from node i to node j. This message represents i's opinion and support about j's all states. Each node considers all the messages from his neighbors and decides the probability of all his states in one iteration. At the end, all messages come to convergence and almost stay changeless. Each node's goodness or badness is determined and estimated marginal probability and is also called belief, $b_i(x_i) \approx P(x_i)$. In a binary state, while the probability of one state is below the fair threshold of 0.5, the node is possibly inclined to the other state. At details, messages are obtained as follows. Messages associated with one edge $edge_{ij}$ are $m_{ij}(x_j)$ and $m_{ji}(x_i)$. Messages are normalized over its recipient, so that all messages point to node j sum to one in every iteration.

$$m_{ij}(x_j) \leftarrow \sum_{x_i \in X} \varphi(x_i)\psi_{ij}(x_i, x_j) \prod_{k \in N(i)-j} m_{ki}(x_i) \tag{3}$$

where node k belongs to the neighboring of node i excerpt for node j. $\varphi(x_i)$ is called the prior potential and the accumulation refers to all states of node i. Formally, $\psi_{ij}(x_i, x_j)$ equals the probability of a node i being in class x_i and its neighbor j in class x_j.

The Algorithm 1 stops if the maximum number of iteration reach or messages converge within threshold. Under most circumstances, the algorithm converges in practice, even though convergence is not guaranteed theoretically. When the algorithm ends, the final beliefs of node are determined as follow:

$$b_i(x_i) = k\varphi(x_i) \prod_{x_j \in N(i)} m_{ij}(x_i) \tag{4}$$

where k is a normalization constant making sure one node's beliefs in all state sum to 1.

Malicious server detection is a long term task, along with new graph coming in, the quantity of client nodes, server nodes and edges increase out of proportion. The incremental LBP in Algorithm 2 is the solution for continuous input and avoiding too many servers node connecting to one clients leading LBP can't work even in insufficient iteration.

Algorithm 1. LBP

Require:
    client-server graph $G = (V, E)$
    compatibility potential ψ
    node's labels L

Ensure:
    class probabilities for each node $i \in V$

1: **for** $e_{ij} \in E$ **do**
2:    $m_{ij} \leftarrow 1, m_{ji} \leftarrow 1$
3: **end for**
4: **repeat**
5:    **for** $e_{ij} \in E$ **do**
6:

$$m_{ij}(x_j) \leftarrow \sum_{x_i \in X} \varphi(x_i)\psi_{ij}(x_i, x_j) \prod_{k \in N(i)-j} m_{ki}(x_i)$$

7:    **end for**
8: **until** messages stop changing in threshold or reaching max iteration number
9: **for** $e_{ij} \in E$ **do**
10:

$$b_i(x_i) = k\varphi(x_i) \prod_{x_j \in N(i)} m_{ij}(x_i)$$

11: **end for**
12: **return** b_i

---

4   Evaluation

The evaluation of this approach is dependent much on obtaining the ground truth of all the servers located all over the world and going deep into the details of actions in the network, that is almost impossible mission. Fortunately we are provided with part of near-ground-truth. The data are provided by a security company. It uses professional technique and tools to detect anomalous behavior and host, keeps a blacklist database for several years. In the following, we describe our data, comparison experiments and then performance results.

4.1   Data Description

First of all, we describe the large data set that we infer the malicious server in the internet. The raw data is captured by network devices that keep trace one intranet with an amount of six thousand of clients from July 14 2015. Every entry contains property such as time, source ip address, destination ip address, port, uri, host and so on. The bipartite graph can be constructed easily according to the source and destination ip address. We extract 7 features for clustering algorithms, such as average length of http uri, average number of parameters in uri, average number of parameters in http cookie, total number of get, total number of post, burst request, total number of requests. In the period, these clients access to the internet server from all around the world with an amount

---

Algorithm 2. Incremental LBP

Require:
    client-server graph stream $GS = (G_1, G_2, \ldots, G_n)$
    compatibility potential ψ
    node's labels L
Ensure:
    class probabilities for each node $i \in V$
1: $b_0 \leftarrow L$
2: **for** $G_i \in GS$ **do**
3:    $b_i \leftarrow LBP(G_i, b_{i-1}, \psi)$
4: **end for**
5:

$$b_i = \begin{cases} anomaly\ if & b_i > 0.5 \\ normal & else \end{cases}$$

6: **return** b_n

---

of 280 thousand by HTTP. The client ip of raw data is hashed to LAN address for information protection and can't be mapped back. In addition, we are provided about 3 thousand of servers with verification information by the company's anomalous server database, which includes one thousand malicious server plus two thousand benign addresses.

4.2 Experiment

We first evaluate the accuracy rate of an individual clustering algorithm for network anomaly detection in Fig. 2. By tuning the weight of each single feature in calculating distance given above, we attain the minimum, maximum and average accuracy rate of different clustering algorithm.

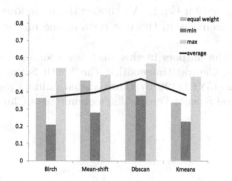

Fig. 2. Average accuracy rate of single recommended clustering algorithm.

It is not hard to find that the accuracy rate of an individual clustering algorithm is far from satisfaction. Thus, we combine the four clustering algorithms and graph, then evaluate the associated prediction accuracy. Here, we set the parameter a (used for Eq. 1 in clustering ensemble) to be 0.2, 0.22, 0.25, 0.27 and 0.3, respectively, and plot the associated accuracy in Fig. 3(a). Compared with the result in Fig. 2, the proposed approach can greatly improve the accuracy.

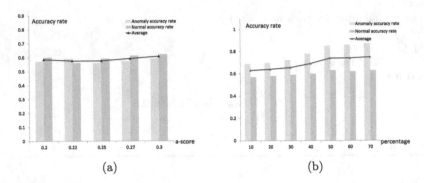

(a)   (b)

Fig. 3. (a) The horizontal axis refers to the experiment number mentioned above. (b) We set the known malicious server nodes with a high score of 0.95. The horizontal axis means the percentage of label data for training.

Ground-truth data are no doubt perfect choice for our algorithm in comparison with the result of cluster ensemble because they're malicious with high credibility. However, it's insufficient for the huge network and only a small quantity is available. Taking the most advantage of the known data and cluster ensemble, we combine both and design the semi-supervised experiment that should be adoptive for maximizing the accuracy in reality. We use part of blacklist database as prior labels and the rest as test data. In Fig. 3(b), we choose randomly from 10% to 70% of malicious servers as label and assign $\varphi_i(x)$ is $\{0.05, 0.95\}$, for the rest $\varphi_i(x)$ is based on Eq. 1. We choose the malicious inference accuracy and benign inference accuracy in the test data as the metric.

Acknowledgement. The authors in this work are sponsored by the Fundamental Research Funds for the Central Universities, the Youth Science and Technology of Foundation of Shanghai (15YF1412600), the Shanghai Sailing Program (17YF1420500) and the National Natural Science Foundation Committee of China under contract no. 61202382.

References

1. Agrawal, S., Agrawal, J.: Survey on anomaly detection using data mining techniques. Procedia Comput. Sci. **60**, 708–713 (2015)
2. Ahmed, T., Oreshkin, B., Coates, M.: Machine learning approaches to network anomaly detection. In: Proceedings of the 2nd USENIX Workshop on Tackling Computer Systems Problems with Machine Learning Techniques, pp. 1–6 (2007)

3. Farid, D., Harbi, N., Rahman, M.: Combining naive bayes and decision tree for adaptive intrusion detection. arXiv preprint arXiv:1005.4496 (2010)
4. Gaddam, S., Phoha, V., Balagani, K.: K-Means+ ID3: a novel method for supervised anomaly detection by cascading K-Means clustering and ID3 decision tree learning methods. TKDE **19**(3), 345–354 (2007)
5. Moore, D., Shannon, C., Brown, D., Voelker, G., Savage, S.: Inferring internet denial-of-service activity. ACM Trans. Comput. Syst. (TOCS) **24**(2), 115–139 (2006)
6. Munson, A., Caruana, R.: Cluster ensembles for network anomaly detection. Technical report, Cornell University (2006)
7. Nachenberg, C., Wilhelm, J., Wright, A., Faloutsos, C.: Polonium: tera-scale graph mining and inference for malware detection (2011)
8. Peddabachigari, S., Abraham, A., Grosan, C., Thomas, J.: Modeling intrusion detection system using hybrid intelligent systems. J. Netw. Comput. Appl. **30**(1), 114–132 (2007)
9. Rayana, S., Akoglu, L.: Collective opinion spam detection: bridging review networks and metadata. In: Proceedings of the 21th ACM SIGKDD International Conference on Knowledge Discovery and Data Mining, pp. 985–994 (2015)
10. Yasami, Y., Mozaffari, S.: A novel unsupervised classification approach for network anomaly detection by k-Means clustering and ID3 decision tree learning methods. J. Supercomput. **53**(1), 231–245 (2010)

TaxiCluster: A Visualization Platform on Clustering Algorithms for Taxi Trajectories

Mingyue Xie$^{(\boxtimes)}$ and Qinpei Zhao

Tongji University, Shanghai, China
836728902@qq.com

Abstract. Taxi is usually considered as the probe of roads in a city. A large amount of taxi GPS mobility data is able to reflect the human mobility and city traffic. The data is described in spatial and temporal form, from which more information can be mined. One kind of the information is related to the basic statistics of the taxi, such as the taxi id, average/min/max speed, travel distance, load or not etc. Other information such as the taxi's trajectory or regions of interest in the city can also be obtained. In this paper, we introduce a *TaxiCluster* to visualize and analyze the taxi data. A procedure on the raw taxi trajectories is introduced in the TaxiCluster.

Keywords: Taxi GPS · Clustering · Visualization · Trajectories

1 Introduction

The GPS-devices have been widely used in people's daily life, especially with the popularity of the smart phones. The data carrying GPS information is so called geo-tagged. The geo-tagged data shows in various formats, for example, a trajectory containing locations in time series, a photo with a location and a time stamp. A taxi traveling every day in the city can generate huge amounts of GPS trajectories. Taxi trajectories hold the information of roads, traffic conditions and even the people in the city. Taxi in a city is a major source of geo-tagged data. Since it travels a lot along different areas and roads in the city, the travel histories from the taxi are quite informative. Therefore, the taxi data has caught the researchers' interest in many areas, such as transportation [1], urban planning, and even advertising [7].

For a large taxi data, two aspects should be considered, which are the data itself and the information hidden. The statistics of the data itself is quite important. The basic statistics that people are interested in, are the starting and stopping location of a taxi for getting regions of interest, the max/min/average speed for estimating the congestion of a road, or the taxi loaded or not for predicting the potential customer pick-up places. The basic statistics are especially useful for obtaining a taxi's profile. Therefore, it is meaningful to develop a visualization platform for demonstrating these basic statistics especially when the

© Springer International Publishing AG 2017
S. Song et al. (Eds.): APWeb-WAIM 2017 Workshops, LNCS 10612, pp. 138–147, 2017.
https://doi.org/10.1007/978-3-319-69781-9_14

data is large. The visualization platform should allow the interaction between the data and the user.

For finding hidden information, data mining methods such as clustering algorithms, as one of the major data mining methods, are usually used. How to perform the clustering algorithms on the taxi data depends on the formats of the data. For a taxi data, a trajectory recording the traveling pattern is a main format. Another useful format is the GPS points, which are not in time sequences, for example, the GPS points for photos taking. For the two formats, the clustering algorithms are designed accordingly. For the clustering of GPS trajectories, how to calculate the similarity of two trajectories is the key problem. Meanwhile, the representatives of the clusters are also an issue. For the GPS points, the clustering algorithms mainly focus on the application, i.e., what is the clustering used to find.

This paper aims to introduce a taxi related geo-tagged data visualization platform. We especially focus on the clustering algorithms employed in the analysis of taxi related data. The clustering for trajectories and starting-and-ending points are discussed.

2   Related Work

There have been a lot of recent research work on taking use of the taxi GPS data from different cities. The trajectories generated by 30,000 taxis from March to May in 2009 and 2010 traveling in urban areas of Beijing [9,11] are used to detect flawed urban planning. In [4], a large scale Singapore taxi dataset consisting of more than 10 million passenger origin-destination GPS points is used to provide useful insight into the city mobility patterns, urban hot-spots, road network usage and general patterns of the crowd movement within the city of Singapore.

The analysis methods for taxi GPS data can be generally classified into two types, which are trajectory-based and GPS point-based. The former one is a challenge task due to the calculation of the trajectory similarities [2,6]. The challenge comes from the fact that the trajectories have semantic features with different direction, speed and combination of routes. Therefore, a set of sub-trajectories are partitioned for getting the total similarity in [6]. On the other hand, clusterings on the GPS points are more classic. Traditional clustering algorithms such as partition-based k-means, density-based DBSCAN and graph-based spectral clustering, are employed in the task. The tricky part is that the GPS points to be processed are usually origin-destination points of passengers or vehicles [3,4,10], photos containing GPS information [5,12]. The purpose of the analysis is to detect events or areas of interest.

3   TaxiCluster

A typical taxi data is shown Table 1. The *GPS ID* is a unique identity number for each GPS data point from the taxi and *Taxi ID* is the sequence number for the taxi. *Longitude* and *Latitude* are the essential values for a location. The

Speed indicates the instant speed of a taxi when its GPS data is collected, while the *Angle* represents the direction of the taxi and the *Time* is in the format of YYYY-MM-DD hh:mm:ss. The *Status* is the only feature that reflects the data is from a taxi because it represents whether the taxi is occupied or unloaded.

Table 1. A taxi data.

GPS ID	Taxi ID	Longitude	Latitude	Speed	Angle	Time	Status
553817420	88685	121.5246	31.2811	30	22	2007-10-12 00:13:34	0
553817421	89289	121.3998	31.1955	26	0	2007-10-12 00:16:34	1
...	...	...	...	...	...	...	...
553817427	3217	121.3441	31.2598	46	45	2007-10-12 08:10:20	0

Since the original data is usually huge, a preprocessing step is needed for removing noises. Duplicated data with same taxi ID and time stamp can be considered as one kind of the noises. A trajectory with less than three points is considered incomplete, which should be removed.

We introduce a platform *TaxiCluster* from three aspects (see Fig. 1), which can be visited from idatatongji.com/taxiClus. The first part is all about the data. The basic statistics on the data is calculated and demonstrated. The second part is about the work on trajectories. We introduce an similarity measure on two trajectories and a clustering algorithm is thereby designed for aggregating the trajectories. The last part is mainly on the clustering for the GPS points.

3.1 Basic Statistics

From the features of a GPS taxi data in Table 1, we give the following definitions.

Definition 1. *A basic GPS data P is a collection of points p with values of latitude, longitude and a time stamp,* $\mathbf{P} = P_1, P_2, ... P_n$, *where n is the size of the data.*

Definition 2. *A GPS trajectory* $\mathbf{T} = P_t$ *is a special case of the basic GPS data, where the time stamps are sequential.*

Definition 3. *Angle of the current data point* P_c *is the intersection angle of the vehicle's direction* $\overrightarrow{P_c P_n}$ *and the north, where* P_n *is the next point. The angle represents the travel direction of the vehicle and should be in the range of [0, 360].*

Definition 4. *Given a complete trajectory, speed of each GPS data point can be calculated by* $dist(P_{ct}, P_{nt})/iDeltat$, *where* $dist(P_{ct}, P_{nt})$ *is the distance between current location and next location.*

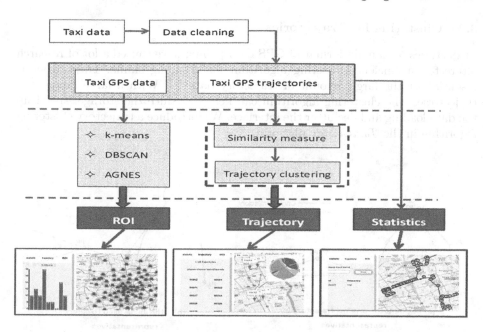

Fig. 1. The framework of the *TaxiCluster*.

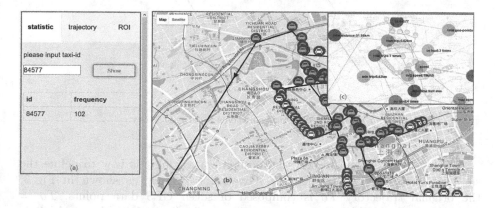

Fig. 2. The basic statistic for a taxi.

Since each taxi has a unique ID, it is possible to give a profile for each taxi. As shown in Fig. 2, the trajectories of a certain taxi can be visualized with the status (occupied or not) shown. For a trajectory (Fig. 2(b)), there are time series of GPS data points. The arrow on each trajectory helps to identify the travel direction. For the selected taxi, the statistics such as average/max/min speed, travel distance for each trajectory, travel distance/time etc. are shown with an information window (Fig. 2(c)).

3.2   Clustering for Trajectories

Trajectories, as a main format of GPS data points, have caused a lot of research interest. An efficient clustering algorithm for grouping similar trajectories is essential for the large amount of trajectory data [6]. With a large amount of trajectories, the clustering algorithm can help to reduce the storage, speed up the data loading and declutter the interface. We introduce a trajectory clustering algorithm in the *TaxiCluster* platform.

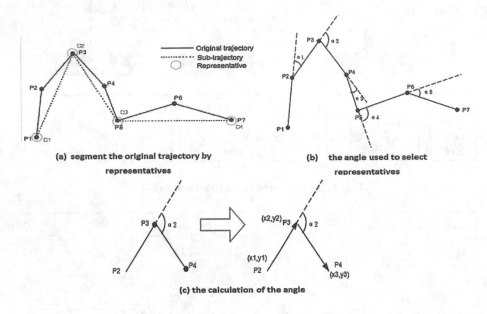

Fig. 3. Select representatives for a trajectory.

In the trajectory clustering, the most important part is how to define the similarity between two trajectories. Given a trajectory as shown in Fig. 3, the original trajectory To is composed of seven GPS data points, $To = \{P1P2P3P4P5P6P7\}$. Since a taxi trajectory is directed by the roads, it is not wise and efficient to calculate the similarity of two trajectories directly by every data point. It is natural to select representatives of the trajectory with the least loss of information. The four data points $\{C1C2C3C4\}$ in Fig. 3(a) are one solution of selecting representatives. For the taxi trajectory, the representatives could be chosen based the travel features of the taxi, for example the significant change of the travel direction. We take the angle as the feature for finding the representatives (see Fig. 3(b)). Taking the $\alpha 2$ as an example, the calculation of it depends on $P2$, $P3$ and $P4$. As shown in Fig. 3(c), the angle $\alpha 2$ is based on vectors $\overrightarrow{P2P3}$ and $\overrightarrow{P3P4}$, which is:

$$cos\alpha2 = \frac{\overrightarrow{P2P3}\overrightarrow{P3P4}}{\|\overrightarrow{P2P3}\|\|\overrightarrow{P3P4}\|}$$

$$= \frac{(x2 - x1)(x3 - x2) + (y2 - y1)(y3 - y2)}{\sqrt{(x2 - x1)^2 + (y2 - y1)^2} * \sqrt{(x3 - x2)^2 + (y3 - y2)^2}} \quad (1)$$

When the calculated angle is large than $\angle 90$, the point is selected as a representative.

After the original trajectories have been reduced with only representatives, the similarity of two trajectories can be calculated by a direction-sensitive Hausdorff distance. Hausdorff distance is typically for measuring two non-empty subsets. In the trajectory case, suppose two trajectories, $T_A = \{P_{A1}, P_{A2}, ..., P_{Am}\}$ and $T_B = \{P_{B1}, P_{B2}, ..., P_{Bn}\}$, the calculation of the direction-sensitive Hausdorff distance is shown in Fig. 4. For the first point P_{A1} in T_A, find the nearest sub-trajectory in T_B, which is TB_{B1}. The shortest distance from P_{A1} to TB_{B1} is d_{AB1}. For the following points, the shortest distances are d_{AB2} to d_{AB5}, which are calculated similarly. A score $Score_{AB}$ for the similarity T_A to T_B is thereby obtained by the average of all the five distances. The same steps are calculated for the $Score_{BA}$, which is the similarity of T_B to T_A. The total similarity between T_A and T_B is taking the average, i.e., $(Score_{AB} + Score_{BA})/2$.

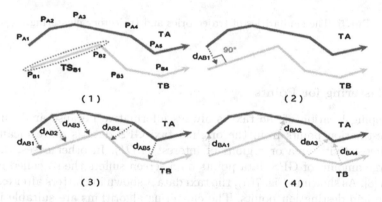

Fig. 4. How to calculate the similarity between two trajectories by the direction-sensitive Hausdorff distance.

With the similarity of two trajectories solved, a trajectory clustering can be implemented to group the similar trajectories together. The steps of the algorithm is as follows:

(1) Initialize each trajectory as one cluster and calculate the pairwise similarity between every two clusters;
(2) Merge the clusters with the largest similarity;
(3) Calculate the pairwise similarities again;
(4) Repeat the step (2) and (3) until a certain number of clusters is reached.

The trajectory clustering is designed for data reduction. With the incremental number of trajectories collected, it is a burden to store, load and visualize the trajectories. After the clustering, the sizes of the original trajectories have been reduced. Meanwhile, the total number of trajectories has also been reduced. In Fig. 5, the pairwise similarities are displayed in a floating window and the clustered trajectories are also shown, where a thicker line connection indicates higher similarity values.

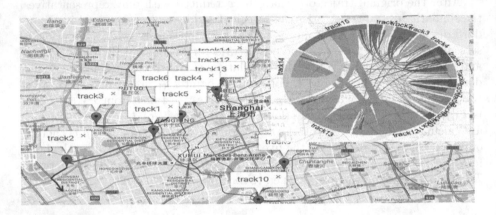

Fig. 5. The similarities of trajectories and trajectory clustering.

3.3   Clustering for Points

The stopping location of the taxi is always informative because analyzing the origin-destination points from the original taxi GPS data is quite meaningful for finding activity area or regions of interest (*ROI*). In other cases, displaying a large amount of GPS data points on a screen suffers the so called clutter problem [8]. As shown in Fig. 7(a), the taxi data is shown in clutters after extracting the origin-destination points. The clustering algorithms are suitable in this case. We implemented three typical clustering algorithms, k-means, DBSCAN and agglomerative nesting (AGNES) for the taxi GPS data, which takes all the origins and destinations from the trajectories. As shown in Fig. 6(a), the three algorithms can be chosen by users. The clustering results are shown in Fig. 6(c). Each cluster is represented by a blue circle with numbers inside. The number indicates the size of the cluster. A graph of bars is shown in Fig. 6(b), which shows the number of clusters in each city districts in the downtown. It is used to analyze the distribution of the functional regions in the city center.

K-means is a partition-based clustering algorithm, it is especially suitable for the data with spherical shape. The algorithm is fast and simple but suffers the initialization problem, which causes the clustering results not the same from every run. The selection of parameter k is an issue, The user needs to

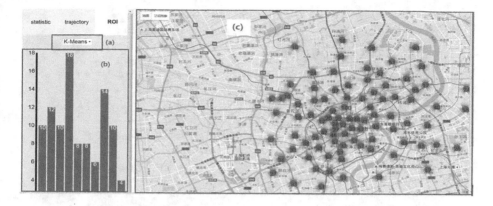

Fig. 6. Clusterings on the GPS points.

determine the number of clusters to be set. DBSCAN groups the data by the density of points. Therefore, the clusters from the algorithm is not restricted to certain shapes. The radius of searching area ϵ and a minimum number of points $MinPts$ required to form a dense region. The tuning of the two parameters brings more difficulty. It is time consuming to tune the two parameters for getting a certain number of clusters k. Therefore, it is difficult to compare the DBSCAN to the clustering algorithms with the same k set. Besides, worse case of the time complexity is $O(N^2)$, which is much higher than the k-means, which is $O(kN)$. AGNES is hierarchical clustering, which takes the merging as a main strategy. The algorithm shares the advantage that it is not restricted to the data shape. Since the algorithm calculates a dendrogram for the whole data, the time complexity is hence higher than the DBSCAN even, which is $O(N^2 \log N)$. Meanwhile, the space complexity is $O(N^2)$.

In the *TaxiCluster*, we intend to demonstrate the three clustering algorithms on the taxi GPS data points. The algorithms are compared to show the different results on the same data in Fig. 7. The number of clusters k for the k-means is set as 100. As shown in Fig. 7(b), the clusters are equally distributed within the city center area and the suburb area. The numbers in the circles indicate the size of the clusters. For DBSCAN, the parameters are set as $\epsilon = 0.2$ and $MinPts = 3$. The clusters from the DBSCAN are those with high density (see Fig. 7(c)). Since the city center is the place with the highest density, the city center has the largest cluster. For the AGNES (see Fig. 7(d)), we stop the merge step when 100 clusters are obtained. The clusters distribute more widely than those from the k-means. Considering the clutter problem, the hierarchical clustering is the most suitable one. For finding the region of interest, the density-based clustering is more suitable. If the efficiency is the only consideration, the k-means then is the best choice.

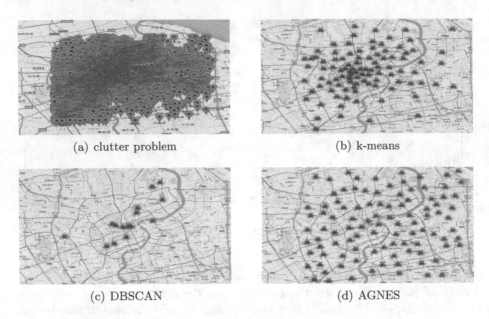

(a) clutter problem

(b) k-means

(c) DBSCAN

(d) AGNES

Fig. 7. Different results from three clustering algorithms on the GPS points.

4 Discussion

Every day, there is a huge amount of taxi data collected. The data contains information on the roads in the city, the travel patterns of passengers and the events happened at a certain place. People may want to visualize the data and get the information from it. It is a trend to develop the visualization platforms for displaying and analyzing large data sets. In the visualization part, the efficiency of the display is one of the problems. For example, when a large of taxi trajectories are queried, the update of the interface should be in seconds. Since the taxi data is displayed on a map, a tricky way is that update the interface according to certain map scale level. A trajectory clustering algorithm can also help in this case, in which lower time complexity is preferred. The clutter problem is another problem in visualization of a large data. With a large size of data and a relatively small screen, the visualization on the data points becomes crowded. Thereby, the clustering algorithms on the data points help to reduce the data size and make the screen neat. In the analysis part, the purpose of the analysis is the key point. Some researchers are interested in the urban planning. A region of interest, subway lines etc. are those they focus on. Some researchers take the taxi data for studying the city traffic condition, finding the easily jammed roads. It is difficult to consider so many purposes in one platform, hence, the introduced *TaxiCluster* is more focused on the visualization part. And the contribution of this paper is the TaxiCluster framework.

5   Conclusion

Taxi's travel data is full of information on a city. The mining and visualization of the data is the focus of the paper. We introduce a platform so called *TaxiCluster* to discuss the clusterings algorithms for both the taxi's trajectories and taxi GPS points. The implementation of a trajectory clustering algorithm is introduced. Meanwhile, three different clustering algorithms for GPS data points are compared and analyzed. TaxiCluster platform combines both the analysis capability and the visualization function. Especially, the design with *D3.js* in the visualization makes the interface in an attractive way.

References

1. Al-Dohuki, S., Kamw, F., Zhao, Y., Ma, C., Wu, Y., Yang, J., Li, X., Zhao, Y., Ye, X., Chen, W.: SemanticTraj: a new approach to interacting with massive taxi trajectories. IEEE Trans. Vis. Comput. Graph. **1**(1) (2016)
2. Evans, M., Oliver, D., Shekhar, S., Harvey, F.: Summarizing trajectories into k-primary corridors: a summary of results. In: ACM SIGSPATIAL IWGS, pp. 454–457 (2012)
3. Guo, D., Zhu, X., Jin, H., Gao, P., Andris, C.: Discovering spatial patterns in origindestination mobility data. Trans. GIS **16**(3), 411–429 (2012)
4. Kumar, D., Wu, H., Lu, Y., Krishnaswamy, S., Palaniswami, M.: Understanding urban mobility via taxi trip clustering. In: IEEE International Conference on Mobile Data Management (2016)
5. Lee, I., Cai, G., Lee, K.: Exploration of geo-tagged photos through data mining approaches. Expert Syst. Appl. **41**(2), 397–405 (2014)
6. Lee, J., Han, J., Whang, K.: Trajectory clustering: a partitiion-and-group framework. In: SIGMOD, p. 593 (2007)
7. Liu, D., Weng, D., Li, Y., Bao, J., Zheng, Y., Qu, H., Wu, Y.: SmartAdP: visual analytics of large-scale taxi trajectories for selecting billboard locations. IEEE Trans. Vis. Comput. Graph. **1**(1) (2016)
8. Zhao, Q., Liao, Z., Li, J., Shi, Y., Tang, Q.: A split smart swap clustering for clutter problem in web mapping system. In: WI, pp. 439–443 (2016)
9. Zhang, J., Qiu, P., Duan, Y., Du, M., Lu, F.: A space-time visualization analysis method for taxi operation in beijing. J. Vis. Lang. Comput. **31**, 1–8 (2015)
10. Zhang, W., Li, S., Pan, G.: Mining the semantics of origin-destination flows using taxi traces. In: ACM Conference on Ubiquitous Computing, pp. 943–949 (2012)
11. Zheng, Y., Liu, Y., Yuan, J., Xie, X.: Urban computing with taxicabs. In: Proceedings of the 13th International Conference on Ubiquitous Computing, pp. 89–98 (2011)
12. Zhou, J., Zhao, Q., Li, H.: Integrating time stamps into discovering the places of interest. In: Huang, D.-S., Jo, K.-H., Wang, L. (eds.) ICIC 2014. LNCS, vol. 8589, pp. 571–580. Springer, Cham (2014). doi:10.1007/978-3-319-09339-0_58

GDMA 2016

Transformation of XML Data Sources for Sequential Path Mining

Ruth McNerlan[1], Yaxin Bi[1(✉)], Guoze Zhao[2], and Bing Han[2]

[1] School of Computing and Mathematics, University of Ulster at Jordanstown, Newtownabbey, Co. Antrim BT37 0QB, UK
y.bi@ulster.ac.uk
[2] State Key Laboratory of Earthquake Dynamics, Institute of Geology, China Earthquake Administration, Beijing 100029, China

Abstract. In recent years XML has become one of the most promising ways to define semi-structured data. Data mining techniques devised for detecting interesting patterns from semi-structure data have also grown in popularity, but carrying out such techniques on XML data can be problematic due to its hierarchical structure. Therefore, it has become necessary to transform XML into flattened, path data, so as to enable data mining to be carried out efficiently. However, problems may arise when the XML tree needs to be reconstructed from the traversal path. There are currently many transformation techniques for XML data, many of which take advantage of its tree-like hierarchical structure; but most of these approaches do not allow the XML tree to be reconstructed from the traversal path. In this paper we propose a new approach to the transformation of XML data into path data. The new approach employs a 5 step transformation process along with a new 'Postorder Sequencing' method of traversing the XML tree. The proposed method, on the one hand, can be seen an efficient and effective way of transforming XML data into collections of paths, and on the other hand enables XML trees to be generated from the traversal paths.

Keywords: XML · Transformation · XPath · Sequential data mining

1 Introduction

The eXtensible Mark-up Language (XML) has established itself in recent years as the de facto language for data exchange on the internet. Since XML is self-describing, it is considered one of the most promising means to define semi-structured data which is expected to be ubiquitous in large volumes from diverse data sources and applications on the web [1]. However, information is encoded differently by different information systems. Therefore, to let these information systems communicate and interoperate, it is necessary to transform XML documents [2]. It can be argued that the pinnacle of XML usefulness is its ability to be transformed into a number of different formats including web pages, databases, knowledge bases and other XML documents. Transformation of XML data is especially important when XML is used as the universal data interchange format among web based applications.

© Springer International Publishing AG 2017
S. Song et al. (Eds.): APWeb-WAIM 2017 Workshops, LNCS 10612, pp. 151–160, 2017.
https://doi.org/10.1007/978-3-319-69781-9_15

Data mining is another field which is growing in popularity with techniques such as association, clustering and classification being used to find similarities and interesting patterns in data. However, the hierarchical, tree-like structure of XML data makes it very difficult to carry out these data mining techniques. Transforming XML data into path data is one way of facilitating data mining on such documents. However, as the result of the data mining will be in the form of path data, it is necessary to have a way of transforming the results into XML as this is a much more effective way of representing data. Therefore, an approach to transforming XML data into path data which enables the generation of XML trees is required.

There are currently many different approaches to transforming XML data. The hierarchical nature of XML lends itself very well to being modelled as a tree-like structure and therefore many approaches to transforming XML involve traversing the XML tree – i.e. – visiting each node of the XML document systematically. There are three main methods of tree traversal – preorder, inorder and postorder – none of which alone permit the tree to be reconstructed from the traversal sequence. Therefore, this paper focuses on the development of a transformation approach for XML data which allows an XML tree to be generated from the traversal path.

2   XML Transformation

There are a number of different approaches to XML transformations that that have been implemented in recent years that attempt to make XML transformations quicker and more efficient. However, there is very limited amount of literature that focuses on developing a transformation approach that allows for the generation of an XML tree from the traversal path. Areas of current research within the topic of XML transformations include transforming between relational databases and XML documents, ways of making XML queries more efficient, and structural join operations in XML queries. Therefore, it can be argued that there is a need for research within the area of XML tree generation from path data.

2.1   Tree Traversal

An XML document can be modelled as an ordered, labelled tree (as illustrated in Fig. 1), with a document node serving as the root node [3]. Each node in the tree corresponds to an element or a value. Values are represented by character data and occur at the leaf nodes. The tree edges represent a relationship between two elements or between an element and a value; and each element can have a list of attribute-value pairs associated with it [4]. The tree structure is considered as the most advantageous data structure in terms of knowledge representation, navigation and access time and therefore many approaches to transforming XML data are based on the tree structure [5].

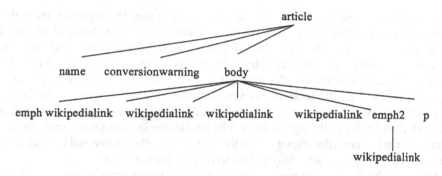

Fig. 1. A sample XML tree

Please note that the first paragraph of a section or subsection is not indented. The first paragraphs that follows a table, figure, equation etc. does not have an indent, either.

Trees are well studied mathematical entities that are used in many ways for organizing and storing information. In all such cases, the information which is represented as terminal nodes (leaves) is retrieved through simple search algorithms that systematically visit every node in the tree. They typically do this in a recursive manner, starting at the root node and descending the tree node by node in the particular order dictated by the algorithm until the search is terminated. The three most commonly used algorithms are, preorder, postorder and inorder with each method producing a different order in which the nodes are visited during the traversal [6]. For example, consider the binary tree in Fig. 2. The preorder traversal visits the root node first and then each child in turn from left to right. The preorder traversal sequence of the tree in Fig. 1 would be A, B, D, E, C, F, G. The inorder traversal method visits the root node in between visiting the left and right subtrees. The inorder traversal sequence of the tree in Fig. 1 would be D, B, E, A, F, C, G. Post order traversal visits the children first before the parent and then visits the root node last. The postorder traversal of the tree in Fig. 1 would be D, E, B, F, G, C, A.

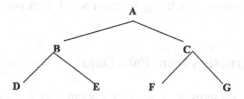

Fig. 2. A binary tree

2.2 Tree Reconstruction

A major limitation of these traversal methods is that it is impossible to reconstruct an XML tree from the traversal path. For example, if ABC is the preorder sequence of a binary tree we can see that A must be the root, but the sequence doesn't tell us if B is the left child of the root or the right child. If ABC is the inorder sequence we don't even know which is at the root node. And if ABC is the postorder sequence, other than knowing that C is at the root we don't know if A is the left or right child and whether B is the left child of A or the right child. We need the inorder sequence and either the preorder or postorder sequence in order to be able to reconstruct the original tree. The inorder sequence will resolve the left-right problem and the preorder or postorder sequence will tell us the roots of the various subtrees. For example, if the inorder sequence is CBA and the preorder sequence is CAB, then we know that C is at the root and both B and A are in the right subtree. The left subtree of the root is empty. Since A comes next in the preorder sequence it is the root of the right subtree and looking at the inorder sequence B is its left child and we have a unique tree [7].

However, XML trees are general trees rather than binary trees as they have nodes with more than two children and there is no sense of left or right. The inorder traversal method is only applicable to binary trees, thus the method of combining the inorder and preorder or postorder traversal sequences to reconstruct the tree does not work for XML trees. Therefore, an alternative approach to traversing XML documents is required that allows the tree to be reconstructed from the traversal path.

2.3 XPath

There are currently many different approaches to transforming XML data including the W3 standard eXtensible Stylesheet Language for Transformations (XSLT) and the XML Query Language (XQuery); and the one thing the majority of these approaches have in common is the use of regular path expressions. Regular path expressions are based on the syntax of the XML Path Language (XPath). They describe traversals through an XML document and consist of one or more steps separated by a slash (/) [8].

The name of the XPath language derives from its most distinctive feature, the path expression, which provides a means of hierarchic addressing of the nodes in an XML tree. XPath gets its name from its use of a path notation for navigating through the hierarchical structure of an XML document. It uses a compact, non-XML syntax to facilitate use of XPath within URIs and XML attribute values. The language provides several kinds of expressions which may be constructed from keywords, symbols, and operands.

3   Tree Reconstruction from Path Data

There are two main problems associated with attempting to generate a tree from a traversal path. These are determining ancestor-descendant relationships and differentiating between identical siblings. Therefore, an approach to XML transformation is required which provides a solution to both these problems and allows an XML tree to be generated from the traversal path.

3.1   Ancestor-Descendant Relationships

XML indexing is one common method that is used to determine ancestor-descendant relationships [9]. XML data can be queried by a combination of value search and structure search. The former includes matching document names, element names/values, and attribute names/values. The latter is usually specified by regular path expressions and is done by examining ancestor-descendant relationship in an XML tree. The main role of indexing XML data is to support both these searches [6].

Another approach to determining ancestor-descendant relationships involves labelling the nodes of an XML tree with a numbering scheme. It was Dietz's numbering scheme that was first to use tree traversal order to determine the ancestor-descendant relationship between any pair of nodes. Dietz uses preorder and postorder traversal order to achieve this [10]. Dietz's approach is based on the proposition that for any 2 given nodes x and y of an XML tree, T, x is an ancestor of y if x occurs before y in the preorder traversal of T and after y in the postorder traversal. Dietz's numbering scheme labels each node with a pair of preorder and postorder numbers, as illustrated in Fig. 3. Examining these numbers allows the ancestor-descendant relationship to be determined.

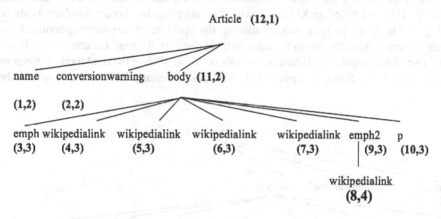

Fig. 3. An XML tree labeled with Dietz's numbering scheme

3.2   Identical Siblings

Along with ancestor-descendant relationships, another problem which is encountered when attempting to reconstruct the XML Tree from the traversal path is identical siblings. For example, if we look again at the example in Fig. 1. A combination of inorder and either preorder or postorder traversal can only be used to reconstruct the binary tree from the traversal path if the tree does not contain any identical siblings – i.e. – any two or more nodes that are the same. Therefore, an approach to XML transformation is needed which solves both problems of ancestor-descendant relationships and identical siblings.

4   A New Approach to XML Transformations

In this study, we propose Postorder Sequencing as a solution to the problem of tree generation. This new approach to XML transformations provides an effective and efficient way to transform XML data into path data and allows XML trees to be generated from the traversal path. We also propose using the Postorder Sequencing method as part of a 5 step approach to the transformation of XML documents.

4.1   Postorder and Level Order Numbering

Given an XML tree are many possible traversal paths that may be navigated through it. Figure 4 depicts an XML tree labelled with a preorder numbering scheme and shows several of the possible paths that can be traversed through the tree. However, none of these paths make it possible to reconstruct the XML tree; more information is needed to be able to determine the exact positioning of each node. We discovered that one solution to this problem is to use a combination of traversal paths to provide the information necessary to enable tree reconstruction. In a similar scheme to Dietz numbering scheme [11], we labeled an XML tree with postorder and level order numbers as shown in Fig. 5 which can be seen as a solution to the problem of tree reconstruction. Level order traversal visits the nodes in order of their depth in the tree, i.e. the root node is at level one, the immediate children of the root node are at level two and their children are at level three etc. Being equipped with both postorder and inorder traversal number

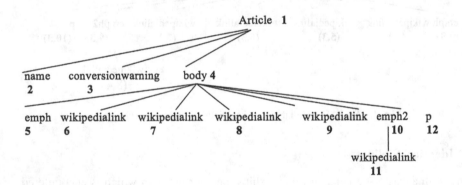

Path 1: 1,2,3,4,5,6,7,8,9,10,11,12 (preorder)
Path 2: 2, 3, 5,6,7,8,9,11,10,12, 4,1 (postorder)
Path 3: 1,2,3,4,5,6,7,8,9, 10,12, 11 (level order)
Path 4: 1, 4, 12, 10, 11, 9, 8, 7, 6, 5, 3, 2 (this is a kind of reverse preorder – the root is visited first, then the rightmost child of the root, then its children starting from the rightmost, then the children, etc.)
Path 5: 12, 11, 10, 9, 8, 7, 6, 5, 4, 3, 2, 1 (this is a kind of reverse postorder –the children are visited before the parents in postorder fashion but the ordering is right to left rather than left to right and the root is visited last)

Fig. 4. Shows a selection of the possible paths through an XML document

patterns when attempting to reconstruct the original XML tree, enables the ancestor-descendant relationship to be determined as the exact position of the node in the tree including depth and horizontal position can be ascertained.

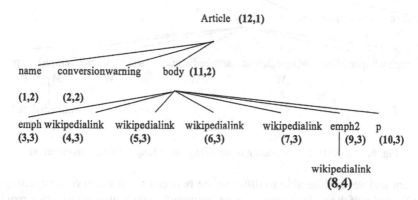

Fig. 5. An XML tree labelled with a postorder and level order numbering scheme

4.2 Postorder Sequencing

We propose postorder sequencing as a new approach to XML transformations which allows for the generation of XML trees as it solves both problems of ancestor-descendant relationships and identical siblings. It can be seen as a development of the level order and postorder numbering scheme but represented in a more effective way.

The work carried out by Wang and Meng's on the sequencing of tree structures for XML indexing [12] can be seen as relevant to the development of an XML transformation approach based on postorder and level order numbering schemes. Their transformation approach involves assigning a 'designator' to each attribute or element of the XML tree. They then map out a preorder sequence of the tree using the designators. Figure 6 shows an XML tree encoded with a designator for each node. Wang and Meng [12] outline how the encoded nodes in a preorder sequence give enough information to reconstruct the tree. For example, the preorder traversal of the tree in Fig. 6 would be A, AN, AC, AB, ABE, ABW, ABW, ABW, ABW, ABM, ABMW, ABP. It is evident that by using this sequence alone the XML tree would be able to be reconstructed.

However, Wang and Meng [12] did not consider the possibility of postorder sequencing. The postorder sequencing of the tree in Fig. 6 would be - NA, CA, EBA, WBA, WBA, WBA, WBA, WEBA, MBA, PBA, BA, A. This sequence allows the tree to be reconstructed as it maps each node from last child to root. It differs from preorder sequencing in that the child nodes are visited before the parent. Postorder sequencing can be seen as a solution to the problem of XML tree generation from path data as it solves both the problems of ancestor-descendant relationships and of identical siblings. The postorder sequence maps out the exact position of the node in the tree determining which node is the parent and which is the child. Some numbering schemes may encounter problems when identical siblings were encountered such as in the tree in Fig. 6. However, the postorder sequencing approach allows there to be many identical siblings in an XML

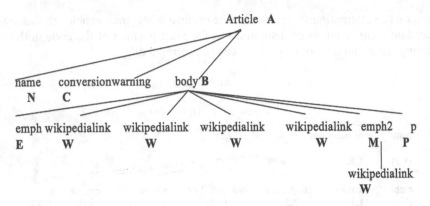

Fig. 6. An XML tree encoding using Wang and Meng's sequencing method

document and yet still be able to differentiate between them when reconstructing the tree as the order of the nodes in a sequence can supply the structural information required to differentiate between identical siblings. Therefore, postorder sequencing can be seen as a solution to the problem of tree reconstruction. This approach can be used to transform XML data into path data for data mining purposes. Once the association, classification or clustering techniques have been carried out, the order of the nodes in the sequence allow for the generation of an XML tree.

Postorder sequencing can be seen as furthering the idea of numbering XML trees with postorder and level order numbers. However, it has the advantage of only having to label the tree with one designator, which, when the sequence path is written out, conveys all the necessary information to reconstruct the tree from the sequence. Postorder sequencing allows all of the child nodes to be collected before the parent node which I believe to be an effective way of grouping XML data.

Postorder sequencing can be considered a novel approach to XML transformations as there is a distinct gap in research when it comes to postorder traversal of XML documents. The majority of indexing or numbering schemes for XML data are based around preorder traversal. e.g., ViST [13], XISS [1], Wang and Meng's Sequencing of Tree Structures for XML Indexing [12], and Tulder's Modified Preorder Tree Traversal [14]. Therefore, there is a need for an evaluation of a postorder traversal approach for XML data in order to determine its usefulness. From my own experiences to date, I consider the postorder sequencing approach to XML transformations to be extremely useful and efficient in transforming XML data as to enable the output to be used to generate trees.

4.3 A Five Step Approach to XML Transformations

We propose a 5 step approach to the transformation of XML data into path data: (1) examine XML document and construct a Document Component Model; (2) construct a tree structure from the XML document; (3) construct DTD for the XML document; (4) label the tree with the designators; (5) transform the XML document into path data using the postorder sequencing approach.

Firstly, the XML document is examined and a Document Component Model is constructed. This is a document that lists all the components of the XML document and acts as a kind of mapping between the XML file and the transformation format. A hierarchical diagram of the XML tree is then constructed which provides a map through the XML document and allows a DTD to be constructed from this which is the third step in the approach. The tree can then be labelled with the node designators which aids the final step of transforming the XML document by traversing through the tree using the postorder sequencing method. This 5 step approach can be seen as encompassing several of the principals of the 'Document Engineering Approach' as outlined by Glushko and McGrath [15].

5 Summary

As XML has become arguably the most popular method of data exchange on the internet, it has become necessary to develop ways of transforming it into other formats such as web pages, databases, and knowledge bases etc. The ability of XML to be transformed into other formats becomes important in an age when data mining techniques such as association, clustering and classification gain popularity. As such techniques are difficult to carry out on XML data it becomes necessary to transform it into path data. One particular approach to the XML transformations involves traversing the tree structure of XML data; and while this proves to be a very effective approach to the transformation of XML data, the majority of tree traversal approaches do not allow an XML tree to be reconstructed from the traversal path. This is mainly because the ancestor descendant relationship is not able to be determined, and further problems may arise when identical siblings are present. Our study discovered that one way of enabling tree reconstruction from path data was through a numbering scheme that allowed both the horizontal position and the depth of each node in the tree to be identified. Postorder sequencing enables this information to be determined and allows an XML tree to be generated from the sequence path. Postorder Sequencing can be seen as an effective and efficient approach to transforming XML data that allows the XML tree to be generated from the path data. The 5 step approach to XML transformations that was implemented proved to be an effective way of ensuring that XML documents are adequately prepared for transformation. It can also be concluded that transformation approaches based on postorder numbering schemes can be seen as being effective in aiding the transformation of XML data.

Acknowledgement. This work is partially supported by the project funded by the National Natural Science Foundation of China (Grant No. 41374077).

References

1. Li, Q., Moon, B.: Indexing and querying XML for regular path expressions. In: Proceedings of the 27th International Conference on Very Large Data Bases, Roma, Italy (2001)

2. Eder, J., Strametz, W.: Composition of XML-transformations. In: Bauknecht, K., Madria, S.K., Pernul, G. (eds.) EC-Web 2001. LNCS, vol. 2115, pp. 71–80. Springer, Heidelberg (2001). doi:10.1007/3-540-44700-8_7

3. Robie, J.: The Tree Structure of XML Queries. Software AG (1999)

4. Rao, P., Moon, B.: PRIX: indexing and querying XML using prufer sequences. In: The 20th International Conference on Data Engineering, Boston, USA, March 2004

5. Miniaoui, S., Forte, M.W.: Data Mining: From Trees to Strings (2004). http://ali2.unil.ch/articles/miniaoui_ICICIS05.pdf

6. Kural, M.: Tree traversal and word order. Linguist. Inq. **36**(3), 367–387 (2005)

7. Rutgers School of Computer Science Website. http://www.cs.rutgers.edu/~kaplan/503/handouts/btreconNgentrees.html

8. Pokornyy, J.: XML Documents Searching and Indexing. Available from Grant Agency of the Czech Republic. www.ksi.mff.cuni.cz/disg/grant/0912.html

9. Toman, K.: Storing and Indexing XML Data (2005). http://www.ksi.mff.cuni.cz/publications?target=file&field=File&id=2540183668933389818

10. Zhang, W., Lui, D., Li, J.: An encoding scheme for indexing XML data. In: IEEE International Conference on e-Technology, e-Commerce and e-Service, 28–31 March 2004, pp. 525–528 (2004)

11. Xing, G.: Indexing XML Data Using Extended Order and Path Index. ACM (2002)

12. Wang, H., Meng, X.: On the sequencing of tree structures for XML indexing. In: Proceedings of the 21st Conference on Data Engineering (2005)

13. Wang, H., Park, S., Fan, W., Yu, P.S.: ViST: a dynamic indexing method for querying XML data by tree structures. In: Proceedings of the 2003 ACM SIGMOD International Conference (2003)

14. Van Tulder, G.: Storing hierarchical data in a database (2003). www.sitepoint.com/print/hierarchical-data-database

15. Glushko, R., McGrath, T.: Document Engineering. MIT Press, Cambridge (2005)

Graph Summarization Based on Attribute-Connected Network

Siqi Liu, Qinpei Zhao, Jiangfeng Li, and Weixiong Rao[✉]

School of Software Engineering, Tongji University, Shanghai, China
lsq52647@126.com, {qinpeizhao,lijf,wxrao}@tongji.edu.cn

Abstract. Techniques to summarize and cluster graphs are important to understand the structure and pattern of large complex networks. State-of-art graph summarization techniques mainly focus on either node attributes or graph topological structure. In this work, we introduce a unified framework based on node attributes and topological structure to support attribute-based summarization. We propose a summarizing method based on virtual links (node attributes) and real links (topological structure) called Greedy Merge (GM) to aggregate similar nodes into k non-overlapping attribute-connected groups. We adopt the Locality Sensitive Hashing (LSH) technique to construct virtual links for high efficiency. Experiments on real datasets indicate that our proposed method GM is both effective and efficient.

1 Introduction

Nowadays, graphs provide information warehouses for complex networks, such as social networks and co-authorship networks in a variety of applications [7]. The effective and efficient identification of the underlying characteristics of these complex networks has become a challenging task. Graph summarization helps users extract and understand the underlying information.

In literature, clustering and aggregation graph summarization are two important learning techniques that are widely used to summarize graphs [2,10,11]. One of the representative clustering methods is called SA-Cluster [2], which proposes to generate k clusters based on attributes and structural similarities. However, these clusters are isolated without considering attribute connectivity. Meanwhile, the solution requires n^2 random walk-based distance summarizations using matrix multiplication for a graph having n nodes. Thus, it is impractical for the SA-cluster to summarize a large graph. Graph aggregation generally provide the "drill-down" and "roll-up" abilities [9] to navigate the summarization with different resolution. The method called k-SNAP releases topological structure constraint for better attributes compression. Bei et al. at [1] proposes an approach called SGVR to aggregate similar nodes into k groups, which considers topological structure and node attributes as two independent parts. However, these methods do not consider relationships between summarized groups.

In this paper, we propose an approach called Greedy Merge (GM) based on virtual links and real links to aggregate nodes that have similar attributes and

© Springer International Publishing AG 2017
S. Song et al. (Eds.): APWeb-WAIM 2017 Workshops, LNCS 10612, pp. 161–171, 2017.
https://doi.org/10.1007/978-3-319-69781-9_16

structure into k groups. Groups that have the same attributes are connected by virtual links and they generate an attribute-connected network. In this network, users can find useful information efficiently according to node attributes. In order to improve efficiency, we construct virtual links by using Locality Sensitive Hashing (LSH) [4]. The proposed method allows users to analyze and visualize graphs from different attributes.

The main contributions of this paper are: (1) Greedy merge (GM) approach is proposed to summarize attribute-connected groups based on node attributes and topological structure. We introduce two types of node relationships, namely, virtual links and real links. LSH is presented to improve the efficiency of constructing virtual links. (2) Extensive experimental evaluation on real worlds demonstrates that our method is more efficient and effective than the previous approaches SA-Cluster and k-SNAP and SGVR method.

The rest of this paper is organized as follows. Section 2 introduces the problem definition. Section 3 describes the GM algorithm. Experimental results are presented in Sect. 4. Section 5 reviews the related work, and Sect. 6 concludes the paper.

2   Problem Statement

In this section, we present notions and problem definitions.

Definition 1. *Graph summarization: Given a static and undirected graph* $G(V,E,\Lambda)$, *attributed graph summarizing is to partition an attributed graph* G *into* k *disjoint attribute-connected groups.*

Definition 2. *Problem definition: Given a general graph model* $G = (V,E,\Lambda)$, *where* V *is the set of nodes,* E *is the set of edges, and* $\Lambda = \{a_1, a_2, ..., a_m\}$ *is the set of* m *attributes associated with nodes in* V *for describing node properties. We assume that node attributes could be numeric or other types. In order to summarization, we transform these numeric attributes into the categorical domains with smaller cardinalities [12]. The domain of attribute* a_i *is* $Dom(a_i)=\{a_{i1}, ..., a_{in_i}\}$ *where* $|Dom(a_i)| = n_i$. *For ease of presentation, we denote* $a_j(v)$ *as the attribute value of node* v *on attribute* a_j.

In this study, we distinguish the topological structure of the original graph from those in the established graph by using similarities of node attributes. The node relationships in the graph are classified into two types of links, namely, real links and virtual links.

Definition 3. *Real link: Given two nodes* u *and* v *in graph* G *and if* $\exists (u, v) \in E$ *is satisfied, then a real link exists between nodes* u *and* v.

Definition 4. *Similarity: Given a set of attributes for two nodes* u, v *and the number of node attributes* m, *for each attribute* $a_i \in \Lambda$. *We define the similarity between nodes* u *and* v *as*

$$sim(u,v) = \sum_{i=1}^{m} \frac{\mid a_i(u) \bigcap a_i(v) \mid}{\mid a_i(u) \bigcup a_i(v) \mid} \tag{1}$$

Definition 5. *Virtual link: Given two nodes u and v, and a minimum threshold* $\alpha \in (0,1)$, *and if sim(u,v)> α, then a virtual link exists between nodes u and v.*

We take an example from DBLP dataset. Figure 1(a) to (d) show our algorithm step by step. In Fig. 1(a), a node represents an author with three attributes that represent the number of publications at the area of database (DB), artificial intelligence (AI) and information retrieval(IR). Authors with ≥ 20 papers are labeled as HP; authors with ≥ 10 and ≤ 20 are labeled as P and authors with ≤ 10 papers are labeled as LP at each research areas. Author 4 and author 6 collaborate on papers, so there is a real link between them. In Fig. 1(b), nodes are connected by real links that generate a graph. Author 5 and author 7 have same attributes LP, LP in DB and IR field and different attribute in AI field. Thus, a virtual link between the two authors is established with the similarity value is 2/3 according to Definition 4. In Fig. 1(c), author 3 and author 7 have virtual links and common neighbour, so they are aggregated into the same group. In Fig. 1(d), according to Definition 1, 9 authors are aggregated into 3 groups.

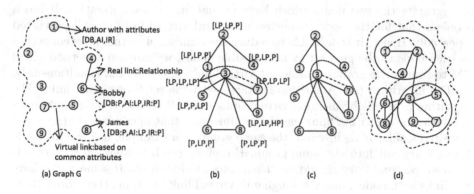

Fig. 1. Graph summarization at each iteration.

Desirable graph summarization should achieve a balance between these properties: (1) Nodes within one group have similar attribute values. (2) Nodes within one group have more real links, while nodes between groups have less real links. (3) For each attribute, the groups having this attribute are connected into an attribute-connected network.

3   GM Algorithm

In this section, we construct virtual links by using LSH for high efficiency (Sect. 3.1), and next we present GM that provides an approximated solution (Sect. 3.2). After that, node adjustment is presented to satisfy the number of groups (Sect. 3.3).

3.1   Locality Sensitive Hashing

Virtual links are constructed according to the similarities between nodes. Low similarities are meaningless for constructing virtual links. For fast search, LSH is a valuable solution that can efficiently find candidate similar nodes based on high similarity, while avoiding unnecessary similarity computations with small-similar nodes. Nodes having high similarities are put into the same bucket by using LSH [4]. Hence, we use LSH to optimize my approach. For this reason, if any two nodes in the same bucket and the similarity of them is more than α, then a virtual link exists between them. For each bucket, virtual links subset S_i is generated. We denote all virtual links set as $S = \{S_1, S_2, ..., S_i\}$. Using LSH to construct virtual links, we only consider the nodes in the same bucket, while avoiding computing any two nodes in the graph.

3.2   Greedy Merge Algorithm

LSH is used to construct virtual links. A naive approach for summarization is to aggregate the two nodes which have virtual link at each iteration. It has a problem that for the node connected to several virtual links, it will be visited repeatedly. Hence, it is desirable to reduce the number of virtual links connected by each node to the greatest extent possible. Thus, we propose a method called GM to summarize graphs considering virtual links and real links simultaneously.

The algorithm (see Algorithm 1) starts with the set S of virtual links. The algorithm proceeds by merging virtual links until all nodes are visited. At each iteration, the algorithm finds the virtual link (u,v) that maximizes sim(u,v), put unvisited similar node u, v into the same group. For example, in Fig. 1(b), node 3 and 7 are put into the same group. However, graph topological structure is omitted. So, real links should be taken into consideration. If some nodes have real links with node u and v belonging to virtual link (u,v), in other words, these nodes are common neighbours of node u and v, they should be aggregated into the group including u and v since these nodes have structural similarities. For example, in Fig. 1(c), the common neighbour of node 3 and 7 is node 9, then they are aggregated into the same group. We denote the common neighbour set as $CN_{u,v}$. If common neighbours of the two nodes are not found, another new group is created (Line 11–12). Whenever there is no virtual link left, the algorithm stops, because this condition means that all nodes are visited. When all nodes are visited, attribute-connected disjoint groups are generated.

In order to improve time efficiency, the minimum number of virtual links are expected. Instead of naively searching for the next best virtual link with maximum similarity, the implementation presented here uses an auxiliary array LinkContrib, whose elements *LinkContrib[i]* are the sets of virtual links V × V with contribution i, i.e. virtual link (v,w)∈ *LinkContrib[i]* and i is the number of common attributes of node u and v. *HighestContrib* is used to hold the biggest i for which *LinkContrib[i]* $\neq 0$. Thus, finding the next best virtual link e = (v,w) and the next best two nodes to aggregate is easy (Line 6). After marking the nodes visited and putting them into the same group and removing virtual link

---

Algorithm 1. Greedy Merge

---

    input : G :a graph, S:virtual links set
    output: Group,which are nodes grouping of G
1 **forall** *each $e=(u,v) \in S$* **do**
2   contrib ← | { $a_i \in \Lambda$: a_i is the common attribute of u and v }|
3   **if** *contrib >0* **then**
4   add e to LinkContrib[contrib]

5 **while** *HighestContrib >0* **do**
6   e ← an edge(u,v) from LinkContrib[HighestContrib]
7   select the unvisited node of u and v
8   unvisited nodes are aggregated into group g_i
9   unvisited nodes are marked visited
10   compute the Common Neighbours set $CN_{u,v}$ of node u and v
11   **if** $CN_{u,v} = \varnothing$ **then**
12   new group are created

13   **if** $CN_{u,v} \neq \varnothing$ **then**
14   unvisited nodes are aggregated into group g_i
15   unvisited nodes are marked visited

16   VirtualLinks ← VirtualLinks −e
17   delete e from LinkContrib[HighestContrib]
18   **forall** *attribute $a_i \in \Lambda$* **do**
19   **forall** *$(u',v') \in S_i, (u',v') \neq (u,v)$* **do**
20   compute i= | {$a_i \in \Lambda$: a_i is the common attribute of u and v }|
21   delete (u',v') from LinkContrib[i]
22   **if** *i> 1* **then**
23   add (u',v') to LinkContrib[i-1]

24   **while** *HighestContrib >0 **and** LinkContrib[HighestContrib] is empty* **do**
25   HighestContrib ← HighestContrib-1

26 *output(Group)*

---

e from *LinkContrib*, the contributions of other virtual links must be updated (Line 18–23). This update can be done efficiently as it is only needed to replace the edges that are in each virtual links subset. For each such virtual link, its corresponding entry in *LinkContrib* can be found in O(1) time if we maintain a pointer to this virtual links entry in *LinkContrib* along with each value of i.

According to Algorithm 1, an attribute-connected network based on node attributes and topological structure is constructed. We analyze the running time of the greedy merge algorithm obtaining an approximate time $O(|V^2| |\Lambda|)$.

3.3 Node Adjustment

After the GM algorithm for grouping nodes based on real links and virtual links, the number of generated groups cannot satisfy expected. A better alternative method is to let users control the sizes of results, then users can manage the summary resolutions.

If the number of groups is more than expected, in order to reduce the isolated nodes, some sparser groups are merged into denser groups because nodes in sparser group are weak connected. At each iteration, the group with the least value of density is considered to be adjusted (see Algorithm 2). The density of each group is defined as the ratio of the number of nodes to the number of

---

Algorithm 2. Node Adjustment

---

    input : *Group*, k:the required number of groups in summary
    output: A summary graph
1 compute density of each group
2 **while** *the number of group* $> k$ **do**
3 | select the least density group g_i
4 | compute g_i with other groups' relationship.
5 | find the group which has strongest relationship with g_i
6 |_ put g_i into this group.

7 **while** *the number of group* $< k$ **do**
8 | select the least density group g_i
9 | obtain the maximum degree node v in the g_i
10 | find the nodes set g_j in g_i have no relationship with v(j> k)
11 |_ g_j are split into new group

12 a summary graph G_{sum}

---

real links. The value of density is larger, the group is more sparser. We find the group that has most real links with the sparsest group, so nodes in the sparsest group should be put into this group. Hence, sparse groups are merged into denser groups. If the number of groups is less than expected, similarly, at each iteration, the sparest group is selected to split into two groups at each time until the number of groups satisfies with the expected number of groups (Line 8). We sort the degree of nodes, then select the maximum degree node and find nodes that have real links with the maximum degree node in this group, so we preserve these nodes and the maximum degree node in the original group, and put the other nodes into new group. The sparest group are split into two denser groups. After the node adjustment, nodes in the same groups are more cohesive and nodes in different groups are incohesive. Then, we analyze the time complexity of node adjustment algorithm. Computing the density of all groups takes the $O(|V|)+O(|E|)$. At each iteration, the time complexity of the node adjustment is the $O(|V^2|)$. Given that $O(|E|)$ is smaller than $O(|V^2|)$, the total time complexity of the node adjustment algorithm is $O(|V^2|)$.

4 Experimental Evaluation

This section presents the experimental results to evaluate the proposed approach and compares to the representative algorithms k-SNAP and SA-Cluster and SGVR in real datasets. Minimum threshold α is set as 0.3. All algorithms are implemented in Java. All experiments were run on Xeon 2.40 GHz cpu (6 cores), 64 GB of RAM, 4 standard 500 GB SSD drive.

4.1 Experimental Datasets and Evaluation Measurements

In this section, we describe the datasets used in our experimental evaluation. We use real datasets to explore the effect of various graph characteristics on the graph summarization. We also use three evaluation criteria to judge the above approaches.

DBLP Dataset. We use the DBLP data from eight research areas of database (DB), artificial intelligence (AI), information retrieval (IR), data mining (DM), network information security (NIS), software engineering (SE), computer network (CN) and computer graphics (CG). For each of the four ares, we collect the publications of some conference in this area. We build a coauthor graph with 17492 authors and their coauthor relationships. By displaying the distribution of the number of author's publications at each research area, publications of each research area presents same distribution. We assign attribute of each aspect called Prolific to each author in the graph indicating whether that the author is prolific, authors with ≥ 20 papers are labeled as highly prolific; authors with ≥ 10 and ≤ 20 are labeled as prolific and authors with ≤ 10 papers are labeled as low prolific.

Enron Email. It is about an email communication network of a company named Enron, where each employee is a node and an email between two persons is an edge. There are total 4256 nodes and 10630 edges. Each node is attached with attributes based on communication email text size. The number of attributes is 142.

We use the density measurement to evaluate the quality of groups generated by above approach, which are showed as followed.

$$density(\{g_i\}_{i=1}^{k}) = \sum_{i=1}^{k} \frac{\mid \{(u,v) \mid u,v \in g_i, (u,v) \in E\} \mid}{\mid V \mid} \tag{2}$$

where k is the number of group, g_i is any generated group by using GM algorithm.

After grouping nodes, the relationships between groups should be considered. We define the participation ratio of the group relationship as

$$\sum_{i=1}^{k}\sum_{j=1}^{k} p_{i,j} = \frac{\mid P_{g_i}(g_j) \mid + \mid P_{g_j}(g_i) \mid}{\mid g_j \mid + \mid g_j \mid} \tag{3}$$

where $P_{g_i}(g_j) = \{u \in g_i$ and $\exists$ v $\in g_j$ s.t. (u,v)$\in$ E$\}$.

After node adjustment, group is not homogenous in node attributes, entropy is used to measure the quality of groups generated.

$$entropy(\{g_i\}_{i=1}^{k}) = \sum_{i=1}^{k} \frac{\mid g_i \mid}{\mid V \mid} entropy(a_i, g_i) \tag{4}$$

where

$$entropy(a_i, g_i) = -\sum_{n=1}^{n_i} p_{ijn} \log_2 p_{ijn} \tag{5}$$

and p_{ijn} is the percentage of nodes in group j which have value a_{in} on attribute a_i. $entropy(\{g_i\}_{i=1}^{k})$ measures the entropy from all attributes over k groups.

4.2   Effective and Efficiency Evaluation

Figure 2 shows the density and participation ratio and entropy respectively among the four methods on the DBLP Dataset when we set the node attribute number m = 8 and the group number k = 25, 50, 75, 100, 125, 150. The density values of GM are higher than other methods. This demonstrates that this method can find densely connected components. On the other hand, when k is small, k-SNAP has a lower density, and the density value decreases quickly when k increases. This is because k-SNAP partitions graphs without considering topological structure. The density of SA-Cluster, SGVR also generally decreases when k increases. The participation ratio measures the relationships between groups. The value of participation ratio is higher, the relationship between the groups is closer. GM has the lowest participation ratio, because GM strikes a balance between topological structure and node attributes. k-SNAP is higher because it only considers node attributes and ignores node connectivity. SGVR and SA-Cluster stand between GM and k-SNAP. The entropy is always 0 for k-SNAP, since it partitions graphs where each group contains nodes with the same attribute value. The entropy of GM decreases when k increases because group homogenous in node attributes is well preserved when the number of node adjustment process is much fewer.

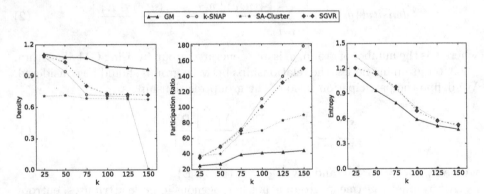

Fig. 2. From left to right: (a) Group density on DBLP Dataset (b) Group participation ratio on DBLP Dataset (c) Group entropy on DBLP Dataset

In this experiment, we evaluate the effectiveness of our method on the enron emails network. From the results, we manually set the number of group with k = 10, 20, 30, 40, 50 on Enron Email. Figure 3(a) and (b) and (c) show the density, participation ratio and entropy respectively when the number of attribute is large. Even though the number of attribute is large, GM outperforms k-SNAP, SA-Cluster and SGVR.

Figure 4(a) to (b) show that the runtime of GM with using LSH is faster than GM without using LSH by 15%. Figure 4(a) indicates the running time (in seconds) of these algorithms on the DBLP dataset when we set the number

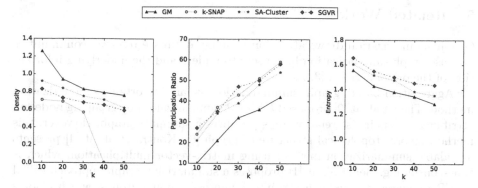

Fig. 3. From left to right: (a) Group density on Enron Emial (b) Group participation ratio on Enron Emial (c) Group entropy on Enron Emial

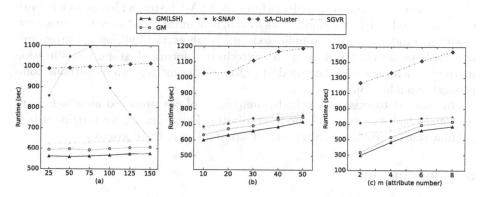

Fig. 4. From left to right: (a) Runtime on DBLP dataset (b) Runtime on Enron Emial (c) Runtime on different number of node attributes

of attributes as 7. GM outperforms k-SNAP and SA-Cluster and SGVR on real datasets using different group numbers. As we use LSH to construct virtual links and store the node similarities in memory for repeated use, the execution time on the calculation of the nodes similarities is reduced. This action improves the performance of GM. SA-Cluster is inefficient because random walk is a time-consuming process. k-SNAP performs better when k is increasing because the number of merge initial group is less with the increasing groups. Figure 4(b) indicates that our algorithm outperforms other graph summarizations on Enron Email when the number of attribute is large. Figure 4(c) indicates the running time (in seconds) of five methods on DBLP dataset when set the number of node attributes m = 2, 4, 6, 8 and the number of groups k = 150. With increasing of the number of node attributes, the runtime of GM is increasing because the time complexity of GM is related to the number of node attributes.

5   Related Work

Graph summarization draws attention to many database research communities because graph summarization help users to understand the underlying information of the original graph [3,5].

Aggregation-based graph summarization is an important summarization method. Tian et al. at [9] proposed k-SNAP summarization by grouping nodes on attributes and pairwise relationships produces a summary graph. However, this method ignores topological structure of graph [6]. Koutra et al. at [4] presents a scalable summarization method using matrix-vector multiplication, which is time-consuming. Bei et al. at [1] proposes a summarizing graph approach called SGVR to aggregate similar nodes into k non-overlapping groups, which ignores graph topological structure.

Clustering graphs is another method to summarize graphs [11]. Cheng et al. at [2] recommend a clustering algorithm called SA-Cluster, which considers both structural and attribute similarities through a unified distance measurement. However, random walk is complicated and is where the bulk of the program's time will be spent. Satuluri et al. introduce a localized graph sparsification method which increases the speed of graph clustering algorithms without compromising quality [8].

In contrast to existing methods, uniqueness of our proposed approach focus on topological structure and node attributes. Furthermore, we provide a new summarization strategy that construct attribute-connected groups.

6   Conclusion

In this paper, we solve the problem of summarizing large graph into attribute-connected groups by considering node attributes and structural similarities. We construct virtual links used by LSH for high efficiency. Furthermore, we propose GM to aggregate nodes in large graph into k non-overlapping attribute-connected groups based on real links (topological structure) and virtual links (node attributes). The experimental results on real datasets demonstrates that our method achieves efficiency and effectiveness.

Acknowledgment. This work is partially sponsored by National Natural Science Foundation of China (Grant Nos. 61572365, 61503286), and Science and Technology Commission of Shanghai Municipality (Grant Nos. 14DZ1118700, 15ZR1443000, 15YF1412600). We also thank the reviewers of this paper for their constructive comments on a previous version of this paper.

References

1. Bei, Y., Lin, Z., Chen, D.: Summarizing scale-free networks based on virtual and real links. Phys. A: Stat. Mech. Appl. **444**, 360–372 (2016)
2. Cheng, H., Zhou, Y., Yu, J.X.: Clustering large attributed graphs: a balance between structural and attribute similarities. TKDD **5**(2), 12 (2011)

3. Chockler, G.V., Melamed, R., Tock, Y., Vitenberg, R.: Constructing scalable overlays for pub-sub with many topics. In: Proceedings of the Twenty-Sixth Annual ACM Symposium on Principles of Distributed Computing, PODC 2007, Portland, Oregon, USA, 12–15 August 2007, pp. 109–118 (2007)
4. Khan, K., Nawaz, W., Lee, Y.: Set-based unified approach for attributed graph summarization. In: 2014 IEEE Fourth International Conference on Big Data and Cloud Computing, BDCloud 2014, Sydney, Australia, 3–5 December 2014, pp. 378–385 (2014)
5. Khan, K., Nawaz, W., Lee, Y.: Lossless graph summarization using dense subgraphs discovery. In: Proceedings of the 9th International Conference on Ubiquitous Information Management and Communication, IMCOM 2015, Bali, Indonesia, 08–10 January 2015, pp. 9:1–9:7 (2015)
6. Mirylenka, K., Cormode, G., Palpanas, T., Srivastava, D.: Conditional heavy hitters: detecting interesting correlations in data streams. VLDB J. **24**(3), 395–414 (2015)
7. Newman, M.E.J.: The structure and function of complex networks. SIAM Rev. **45**(2), 167–256 (2003)
8. Satuluri, V., Parthasarathy, S., Ruan, Y.: Local graph sparsification for scalable clustering. In: Proceedings of the ACM SIGMOD International Conference on Management of Data, SIGMOD 2011, Athens, Greece, 12–16 June 2011, pp. 721–732 (2011)
9. Tian, Y., Hankins, R.A., Patel, J.M.: Efficient aggregation for graph summarization. In: Proceedings of the ACM SIGMOD International Conference on Management of Data, SIGMOD 2008, Vancouver, BC, Canada, 10–12 June 2008, pp. 567–580 (2008)
10. Wu, A.Y., Garland, M., Han, J.: Mining scale-free networks using geodesic clustering. In: Proceedings of the Tenth ACM SIGKDD International Conference on Knowledge Discovery and Data Mining, Seattle, Washington, USA, 22–25 August 2004, pp. 719–724 (2004)
11. Xu, X., Yuruk, N., Feng, Z., Schweiger, T.A.J.: SCAN: a structural clustering algorithm for networks. In: Proceedings of the 13th ACM SIGKDD International Conference on Knowledge Discovery and Data Mining, San Jose, California, USA, 12–15 August 2007, pp. 824–833 (2007)
12. Zhang, N., Tian, Y., Patel, J.M.: Discovery-driven graph summarization. In: Proceedings of the 26th International Conference on Data Engineering, ICDE 2010, 1–6 March 2010, Long Beach, California, USA, pp. 880–891 (2010)

Timely Detection of Temporal Relations in RDF Stream Processing Scenario

Xuanxing Yang[1,3], Guozheng Rao[1,3](✉), and Zhiyong Feng[2,3]

[1] School of Computer Science and Technology, Tianjin University,
Tianjin 300350, People's Republic of China
rgz@tju.edu.cn
[2] School of Computer Software, Tianjin University,
Tianjin 300350, People's Republic of China
[3] Tianjin Key Laboratory of Cognitive Computing and Application,
Tianjin 300350, People's Republic of China

Abstract. Windows, taken as snapshots of an RDF stream, are often applied to effectively characterize the semantics of the stream in RDF stream processing (RSP). Since some temporal relations among RDF triples in a window are possibly missing, however, windows could merely characterize the whole semantics of an RDF stream in a rough way. It is interesting to explore the temporal relations between RDF triples in RDF streams. In this paper, we extend continuous queries by introducing some important temporal relations to characterize inner connections among RDF triples, such as preorder relations between time points and time intervals. Furthermore, we demonstrate that those temporal relations are useful and effective in a traffic application.

Keywords: RDF stream processing · Temporal RDF · Temporal relation

1 Introduction

The Web is highly dynamic, recently with the development of a wide range of time-aware applications such as traffic monitoring systems, smart cities and mobile applications, there has been an emergence of real-time information available on the Web, in the form of continuous data streams. Compared with the conventional relational data, these data streams are collections of rather fine-grained, unstructured, incomplete pieces of information, and most of them are valid at a short term of time. For example, sensor readings, GPS information and tweets on the social media. In order to follow and leverage the dynamic information carried by these streams and further support high-level decisions, efforts have been made to integrate multiple data streams of interest, as well as background knowledge, for querying and reasoning processes. However, the

This work is supported by the programs of the National Natural Science Foundation of China (61373165).

S. Song et al. (Eds.): APWeb-WAIM 2017 Workshops, LNCS 10612, pp. 172–182, 2017.
https://doi.org/10.1007/978-3-319-69781-9_17

continuous nature of data streams have determined that conventional one-time query strategy will no longer be useful, which gives birth to a new data processing model, namely stream processing [1,2].

The data streams are heterogeneous, both in value dimension and time dimension. On one hand, the values of streaming data range from structured values (e.g. sensor readings, GPS data), to unstructured texts (e.g. blogs and tweets on social media), which makes the relatedness between different data streams complicate to manage, not to mention when taking into consideration the background knowledge. On the other hand, the time annotations of streaming data are extremely diverse, some statements are valid over a time interval (e.g. temperature readings), and some are statements of instaneous events of interest (e.g. overspeed detections from road sensor). The diversity of time annotations makes temporal relations between streaming data, which is of crucial significance, hard to quantify and manage, and inevitably leads to the low performance when querying patterns with complex temporal constraints.

Therefore, in order to integrate different data streams and background knowledge, without losing the mutual relatedness both in value dimension and time dimension, it is necessary to lift data streams into semantic level. The Resource Description Framework (RDF) [7,8] and SPARQL [9,10], as the W3C recommendations for representing and querying semantic information on the Web, offer a novel solution for seamless integrating and unified querying processes over multiple data streams, as well as background knowledge, therefore has gained considerable attention from researchers in stream processing area. Following the work of well-established Continuous Query Language [3,4], several efforts [11–13] have been made to extend both RDF and SPARQL for streams and continuous processing, namely RDF stream processing (RSP). Although these works have realized continuously querying functionalities over streaming RDF data, and provided alternative but equivalent semantics on basis of SPARQL language, to date none of these works have addressed the problem of representing, updating and querying the mutual temporal relations between streaming RDF triples, in a fine-grained time dimension.

According to Gutierrez's work about temporal RDF graph [5,6], it requires 4 times more RDF triples to represent time annotation of RDF statement in general RDF graph, which leads to serious scalable problems both in memory consumption and retrieval efficiency. Therefore, existing RSP systems remove the time annotations from RDF stream elements and only load the RDF part of RDF stream elements into memory, as it is shown in Fig. 1(a). This RDF graph contains RDF part of RDF stream elements that are valid during the last time window, as well as the background knowledge: (Redlight1, LocatesAt, RoadA). With these RDF statements, we are able to analyze the traffic condition on RoadA during the last time period. However, the elimination of time annotation makes it impossible to determine the temporal relations between RDF stream elements, which results in a vague and imprecise semantics of RDF stream. For example, we cannot determine if the Car123 ran the red light on RoadA, since

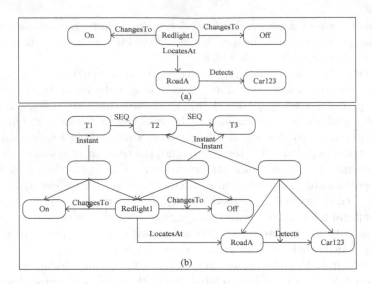

Fig. 1. (a) RDF stream snapshot without time annotation; (b) RDF stream snapshot with time annotation and temporal relations between RDF statements.

both conditions have the same RDF graph representation if we remove the time annotations from RDF stream elements.

In fact, the extra time dimension has changed RDF stream from a bag-of-knowledge to an event-based perspective. The temporal information is an essential composing part of RDF stream semantics that needs to be captured precisely. In order to combine temporal information and RDF statements together, it is necessary to introduce both time and temporal relations into RDF graph, as it is shown in Fig. 1(b). The time annotations of RDF stream elements are represented using RDF reification vocabulary, and the temporal relations between time nodes (denoted by T1, T2, T3) are represented with a pre-defined RDF property (namely SEQ). From this temporal RDF graph, we can infer that the Car123 ran the red light during the last time period, since the time point when Car 123 is detected on RoadA falls in the time interval when the red light of RoadA is on. In this approach, the complex temporal constraint on RDF statements is decomposed into a set of temporal relations and represented with RDF vocabularies, therefore, the temporal constraint can be represented with RDF graph patterns, using SPARQL syntax only, which guarantees the interoperability between temporal constraints and RDF graph patterns, and will benefit from the well-established, scalable support of RDF storage and query engines.

In this paper, we propose some temporal relations between RDF statements. With these temporal relations as primitives, we are able to represent and query arbitrary temporal constraints on RDF stream elements. Our contributions are:

– The analysis for features of RDF stream in time dimension.

- The formal definition of a set of temporal relations customized for RSP scenario, enabling representing arbitrary temporal relations between time points and time intervals.
- A solution for adding these temporal relations without querying the RDF stream.

2   Introducing Temporal Relations into RDF

Generally speaking, a time-annotated RDF triple can be considered as an event statement. The RDF triple states the content of event and the time annotation states the valid time of event. Both time annotation and RDF statement are composing parts of the semantics of time-annotated RDF triple, therefore the temporal correlation between events should be represented explicitly. Although there have been efforts [5,6] to introduce time annotation of RDF statements into RDF graph using RDF reification vocabulary, currently there is no work about extending RDF with temporal relations between RDF statements. We will next discuss the main issues arise when defining the temporal relations.

Versioning vs. Time Labeling. The mechanisms for adding time dimensions to non-temporal RDF graphs are twofold, namely *labeling* and *versioning* [5]. The former approach is to label the changed elements with time annotations, and the latter is to maintain a snapshot for the each state of the system. For example, when new changes occur, a new version is created and the previous version is removed from memory or stored somewhere else. Although both approaches are equivalent, it is more suitable to apply time labeling than versioning in RSP scenario. On the one hand, the streaming data are updated with high frequency, and it is not feasible to create a new graph for every update, in terms of large throughput of RDF streams. On the other hand, the versioning approach separates stream into snapshots, which makes it not suitable for retrieving pattern that varies over a continuous time interval.

Time point vs. Time interval. The next problem arises when enriching time label with RDF triples: to choose time point or time interval as the primitive representation of time. Primarily, both representations are equivalent, a time interval can be seamlessly converted into point-based representation, and vice versa. However, we believe the point-based representation is more suitable in multiple-stream scenario. Firstly, the interval-based representation of time is based on the spirit that every temporal statement is decomposable [14], and a time interval is always decomposable. However, compared with persistent, historical information, the RDF stream elements are short-termed, fragmented information, and our goal is to detect temporal relations between them, therefore point-based representation is more suitable. Secondly, not all temporal statements can be determined in advance, there are time intervals determined by point-based statements from different RDF streams. For example, the route information of one vehicle may be determined by road sensors with different locations, which will can only be detected in a centralized processing system.

Thirdly, point-based representation satisfies the real-time nature of streaming information, since an interval-based statement will be delivered only when the last part of it is detected, which will eliminate the possibility for starting point of it to relate with other statements, since others may be deleted for memory consumption. Thus, in this paper we will use time points as the primitive representation of time, and convert interval-based events into a set of point-based events.

Length of Temporal Relation. The temporal relations between two time points are simple to define, namely in the form: X *before* Y, X *after* Y, and X *equal* Y. Among these three temporal relations, *before* and *after* are used for relating two distinct time points together. However, without the restriction for the length of time interval between two time points, these temporal relations will combine time points that spans too long to change their original semantics. We denote these two temporal relations by *unbounded* temporal relations. In RSP scenario, some events occur repeatedly, the vague semantics of unbounded temporal relation will make query results unpredictable.

For example, in the traffic monitoring system, we use the continuous query below to detect the cars that ran the red light.

REGISTER   QUERY   CarsRanTheRedLight
SELECT   ?car
FROM STREAM   <http://streams.org/roadsensors.trdf>
[**RANGE**   30m   **STEP**   5m]
WHERE   { ?roadsensor   :detects   ?car t_1.
?redlight   :changesTo   :on t_2.
?redlight   :changesTo   :off t_3.
?redlight   :locatesAt   ?roadsensor}
FILTER ($t_2 < t_3$ **AND** $t_2 < t_1$ **AND** $t_1 < t_3$)

Intuitively, this query determines the temporal order of three events, ensuring the time point when car was detected on the road falls in the time interval when the red light was on. However, consider the example data below:

{(:redlight1 :changesTo :on):[0], (:roadA :detects :car1):[1],
(:redlight1 :changesTo :off):[2], (:roadA :detects :car2):[3],
(:redlight1 :changesTo :on):[4], (:redlight1 :changesTo :off):[5],
(:redlight1 :locatesAt :roadA)}

Under this condition, the query answer will be (?car→:car1) and (?car→:car2). Obviously :car2 did not run the red light, it is in the answer because it falls in the interval, that is determined by the first time :redlight1 changes to :on and the second time :redlight1 changes to :off. The original semantics off temporal relation "$t_2 < t_3$" was to identify the time interval when the red light was on, however, without the restriction for interval length, this temporal relation relates two time points spans too long to change the original semantics of it.

Therefore, when defining a temporal relation, it is necessary to specify the length of it. However, in stream processing scenarios, the temporal RDF statements represent real-time events, these events show strong randomness in time dimension and the length of temporal relation cannot be determined. Under this condition, the length of temporal relation can not be measured with timestamp values, but with temporal order, i.e., the more events occurred during a time interval, this time interval is considered to span longer. Therefore, we propose a way to define temporal relation according to the temporal order of RDF triple patterns, which will be illustrated in the next section.

3 Temporal Relation

In this section we will illustrate the formal semantics of the definitions for a set of primitive temporal relations, and how to combine them to represent complex temporal constraints.

3.1 Basic Definitions

In this paper we will work with the point-based temporal domain for defining our data model, and represent interval-based RDF statement with a pair of point-based RDF statements. Consider time as a discrete, linearly ordered domain, we give our formal definition of RDF stream.

Definition 1. *RDF Stream.*

1. *Time domain is a set of natural numbers, $Time = \{t | t \in \mathbb{N}\}$.*
2. *A temporal RDF statement is an RDF triple with a timestamp (a natural number). We will use the notation $(a, b, c) : [t]$.*
3. *RDF stream is a set of temporal RDF statements, $Stream = \{(a, b, c) : [t]\}$.*

Now, we give the definition of our primitive temporal relations.

Definition 2. *Temporal Relations.*
Denote RDF triple patterns with p_1, p_2, p_3.
$Equal(p_1, p_2) = \{t_1 | \exists \mu_1, \mu_2.\mu_1(p_1) : [t_1], \mu_2(p_2) : [t_2] \in Stream \wedge \mu_1 \cup \mu_2 \neq \emptyset \wedge t_1 = t_2\}.$

$Forward(p_1, p_2) = \{(t_1, t_2) | \exists \mu_1, \mu_2.\mu_1(p_1) : [t_1], \mu_2(p_2) : [t_2] \in Stream \wedge \mu_1 \cup \mu_2 \neq \emptyset \wedge t_1 < t_2 \wedge (\forall \mu_3.\mu_3(p_2) : [t_3] \in Stream \wedge \mu_1 \cup \mu_3 \neq \emptyset \rightarrow t_2 - t_1 \leq t_3 - t_1)\}.$

$Backward(p_1, p_2) = \{(t_1, t_2) | \exists \mu_1, \mu_2.\mu_1(p_1) : [t_1], \mu_2(p_2) : [t_2] \in Stream \wedge \mu_1 \cup \mu_2 \neq \emptyset \wedge t_1 > t_2 \wedge (\forall \mu_3.\mu_3(p_2) : [t_3] \in Stream \wedge \mu_1 \cup \mu_3 \neq \emptyset \rightarrow t_1 - t_2 \leq t_1 - t_3)\}.$

$Contains(p_1, p_2, p_3) = \{(t_1, t_3), (t_2, t_3) | (t_1, t_2) \in Forward(p_1, p_2) \wedge \exists \mu.\mu(p_3) : [t_3] \in Stream \wedge t_1 < t_3 \wedge t_3 < t_2\}.$

Generally, our strategy for defining temporal relations is to guarantee that the temporal relations is built only between the nearest time points, that satisfies the temporal order of triple patterns. Therefore we define *Forward* and *Backward* to search for the nearest time point, that exist temporal RDF triples satisfying the triple pattern, from two directions. Note that these two relations can also be used to identify the time intervals of interval-based statements, since the starting and ending points of a certain interval-based statement will have at least one common URI.

Table 1. Pictorial example of overlapping time intervals

1	2	3	4	5	6	7	8	9
X_1	X_1	X_1	X_1	X_1	X_1			
		X_2	X_2	X_2	X_2	X_2	X_2	X_2
			A					

With relations *Equal, Forward, Backward*, we are able to define arbitrary binary temporal relations between point-based statements. However, they are not enough to define temporal relations between interval-based statements. As it is shown in Table 1, we cannot determine point-based statement A during which interval, since use *Backward* and *Forward* relations to search for the nearest starting and ending points of interval-based statement X, we will get two distinct pointed-based statements, namely X_2 at time point 3 and X_1 at time point 6. This is because same type of statements cannot be distinguished with single triple pattern, therefore, in situations where same type of interval-based statements will overlap, we need to use *Forward* or *Backward* relations to identify the interval-based statements in advance. Thus, we have introduced relation *Contains* to determine if a point-based statement with triple pattern p_3, falls in the intervals identified by relation *Forward*(p_1, p_2). Note that we will represent relation *Contains* with two binary relations.

Now with these four temporal relations as primitives, we are able to represent arbitrary binary temporal relations between point and point, point and interval, interval and interval, respectively. All of these definitions strictly restrict the number of temporal relations, and more importantly, have a clear and precise semantics, are suitable for detecting complex events in stream processing scenario.

3.2   Expressivity

In this section we show the expressivity of these primitive temporal relations, by using them to represent arbitrary binary temporal relations between time points and time intervals, including the 13 kinds of relations between time intervals defined in Allen's interval logic [14], with relations *before* and *after* replaced with relations *Forward* and *Backward*, respectively. As it is shown in Table 2, for

Table 2. Temporal relations between time points and time intervals

Relation	Point-based Relations	Formal Representation of Temporal Constraints	Pictorial Example
A forward B	A forward B	$\exists t_1, t_2. (t_1, t_2) \in Forward(A, B)$	A B
A backward B	A backward B	$\exists t_1, t_2. (t_1, t_2) \in Backward(A, B)$	B A
A equal B	A equal B	$\exists t.t \in Equal(A, B)$	A B
A forward X	A forward X_s	$\exists t_1, t_2. (t_1, t_2) \in Forward(A, X_s)$	A XXX
A backward X	A backward X_e	$\exists t_1, t_2. (t_1, t_2) \in Backward(A, X_e)$	XXX A
A start X	A equal X_s	$\exists t.t \in Equal(A, X_s)$	A XXX
A end X	A equal X_e	$\exists t.t \in Equal(A, X_e)$	A XXX
A during X	A during X_s A during X_e	$\exists t_1, t_2, t_3. (t_1, t_2, t_3) \in Contains(X_s, X_e, A)$	A XXX
X forward Y	X_e forward Y_s	$\exists t_1, t_2. (t_1, t_2) \in Forward(X_e, Y_s)$	XXX YYY
X meet Y	X_e meet Y_s	$\exists t.t \in Equal(X_e, Y_s)$	XXX YYY
X equal Y	X_s equal Y_s X_e equal Y_e X_s forward1 X_e Y_s forward2 Y_e	$\exists t_1, t_2. t_1 \in Equal(X_s, Y_s)$ $\wedge t_2 \in Equal(X_e, Y_e)$ $\wedge (t_1, t_2) \in Forward(X_s, X_e)$ $\wedge (t_3, t_4) \in Forward(Y_s, Y_e)$	XXX YYY
X overlap Y	Y_s during1 X_s Y_s during1 X_e X_e during2 Y_s X_e during2 Y_e X_s forward1 X_e Y_s forward2 Y_e	$\exists t_1, t_2, t_3, t_4.$ $(t_1, t_2, t_3) \in Contains(X_s, X_e, Y_s)$ $\wedge (t_3, t_4, t_2) \in Contains(Y_s, Y_e, X_e)$ $\wedge (t_1, t_2) \in Forward(X_s, X_e)$ $\wedge (t_3, t_4) \in Forward(Y_s, Y_e)$	XXX YYY
X during Y	X_s during1 Y_s X_s during1 Y_e X_e during2 Y_s X_e during2 Y_e X_s forward1 X_e Y_s forward2 Y_e	$\exists t_1, t_2, t_3, t_4.$ $(t_3, t_4, t_1) \in Contains(Y_s, Y_c, X_s)$ $\wedge (t_3, t_4, t_2) \in Contains(Y_s, Y_e, X_e)$ $\wedge (t_1, t_2) \in Forward(X_s, X_e)$ $\wedge (t_3, t_4) \in Forward(Y_s, Y_e)$	XXX YYYYY
X start Y	X_s equal Y_s X_e during Y_s X_e during Y_e X_s forward1 X_e Y_s forward2 Y_e	$\exists t_1, t_2, t_3, t_4.$ $t_1 \in Equal(X_s, Y_s)$ $\wedge (t_3, t_4, t_2) \in Contains(Y_s, Y_e, X_e)$ $\wedge (t_1, t_2) \in Forward(X_s, X_e)$ $\wedge (t_3, t_4) \in Forward(Y_s, Y_e)$	XXX YYYYY
X end Y	X_e equal Y_e X_s during Y_s X_s during Y_e X_s forward1 X_e Y_s forward2 Y_e	$\exists t_1, t_2, t_3, t_4.$ $t_2 \in Equal(X_e, Y_e)$ $\wedge (t_3, t_4, t_1) \in Contains(Y_s, Y_e, X_s)$ $\wedge (t_1, t_2) \in Forward(X_s, X_e)$ $\wedge (t_3, t_4) \in Forward(Y_s, Y_e)$	XXX YYYYY

the sake of space, we have omitted the reverse version of relations between time intervals, i.e., relations Y meet X, Y overlap X, Y during X, Y start X, and Y end X, are omitted.

Note that A and B are point-based triple patterns, X_s and X_e, Y_s and Y_e are point-based triple patterns of the starting points and ending points of interval-based statements X and Y, respectively. In this way, we are able to convert temporal relations between time intervals into a set of point-based temporal relations, without representing intervals as primitives.

4   Implementation

In this section we offer a solution for adding temporal relations without querying the RDF stream.

4.1   History of Triple Patterns

From event-based perspective, a time-annotated RDF triple can be considered as an event, and the event type is determined together by the subject, predicate and object of this triple. Therefore, if we collect all the time instants of every distinct subject, predicate, object in RDF stream, we are able to determine all the time instants of a certain triple pattern, denoted by the *History* of this triple pattern. Next we illustrate our solution through formal method.

Definition 3. *History of distinct subject, predicate, object in Stream.*
$SHistory(s) = \{t|\forall \mu.(\mu(s), b, c) : [t] \in Stream\}$.
$PHistory(p) = \{t|\forall \mu.(a, \mu(p), c) : [t] \in Stream\}$.
$OHistory(o) = \{t|\forall \mu.(a, b, \mu(o)) : [t] \in Stream\}$.

In this way, we define history of every distinct subject, predicate, object in RDF stream. Note that for all $v \in V$, there is always a mapping from v to every distinct subject, predicate, object in stream. Therefore, we have: $\forall v \in V, SHistory(v) = PHistory(v) = OHistory(v) = \{t|(a, b, c) : [t] \in Stream\}$. Further, we give the definition of history of triple pattern.

Definition 4. *History of triple pattern*
For a triple pattern tp=(s,p,o), define
$History(tp) = \{t|t \in SHistory(s) \cap PHistory(p) \cap OHistory(o)\}$.

4.2   Redefining Temporal Relations

Now that we are able to determine the results of temporal relations that two triple patterns share no common variables, simply by comparing the *History* of these two triple patterns. However, in some definition of temporal relations, the two triple patterns are related through common variables or background knowledge, as it is shown in Fig. 2.

Our strategy is to redefine these temporal relations by grounding the related variables. In the case of Fig. 3(a), we will rewrite the temporal relation $Forward((?A, :ChangesTo :On), (?A, :ChangesTo, :Off))$ by replacing the variable ?A with every distinct subject occurred in stream. In the case of

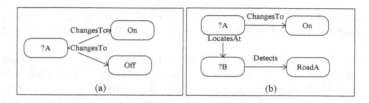

Fig. 2. Examples when two triple patterns are related: (a) share same variable; (b) related with background knowledge.

Fig. 3(b), the original temporal relation is: *Forward*((?A, :ChangesTo, :On), (?B, :Detects, ?C)), and these two triple patterns are related with background knowledge: (?A, LocatesAt, ?B), thus, we need to execute query (?A, LocatesAt, ?B) on the background knowledge, and fill the answers into the triple patterns. For example, if we get mappings μ_1(?A→:redlight1, ?B→:roadA) and μ_2(?A→?redlight2, ?B→:roadB), we will redefine this temporal relation into: *Forward*((:redlight1, :ChangesTo, :On), (:roadA, :Detects, ?C)), and *Forward*((:redlight2, :ChangesTo, :On), (:roadB, :Detects, ?C)), both the two relations will be represented with the same RDF property since they have the same semantics.

With this approach, we are able to determine the results of all the temporal relations through comparing the *History* of triple patterns, and adding these temporal relations into the temporal RDF graph without querying the RDF stream.

5   Conclusions

In this paper, we analyzed the distinguishing features of RDF stream in time dimension, and proposed the definition for a set of binary temporal relations to represent arbitrary temporal relations between time points and time intervals, enabling further exploration of temporal information in RSP scenario.

However, from data processing point of view, state-of-art stream processing systems are still limited by memory bound. That is, for the sake of memory space, RSP systems cannot maintain too many streaming data in memory, which eliminates the possibility for detecting compound temporal relations that spans for a long time interval. Our next work will try to realize a RSP system, that remove the temporal RDF triples not just according the predefined time window function, but also take into consideration of the existing temporal relations in time window.

References

1. Babcock, B., Babu, S., Datar, M., et al.: Models and issues in data stream systems. In: ACM SIGMOD-SIGACT-SIGART Symposium on Principles of Database Systems, pp. 1–16 (2002)

2. Margara, A., Urbani, J., Harmelen, F.V., et al.: Streaming the web: reasoning over dynamic data. Web Semant. Sci. Serv. Agents World Wide Web **25**(1), 24–44 (2014)

3. Arasu, A., Babu, S., Widom, J.: CQL: a language for continuous queries over streams and relations. In: Lausen, G., Suciu, D. (eds.) DBPL 2003. LNCS, vol. 2921, pp. 1–19. Springer, Heidelberg (2004). doi:10.1007/978-3-540-24607-7_1

4. Arasu, A., Babu, S., Widom, J.: The CQL continuous query language: semantic foundations and query execution. VLDB J. 121–142 (2006)

5. Gutierrez, C., Hurtado, C., Vaisman, A.: Temporal RDF. In: Gómez-Pérez, A., Euzenat, J. (eds.) ESWC 2005. LNCS, vol. 3532, pp. 93–107. Springer, Heidelberg (2005). doi:10.1007/11431053_7

6. Gutierrez, C., Hurtado, C., Vaisman, A.: Introducing time into RDF. IEEE Trans. Knowl. Data Eng. **19**(2) (2007)

7. Hayes, P.J., Patel-Schneider, P.F.: RDF Semantics 1.1. W3C Recommendation (2014)

8. Klyne, G., Carroll, J.J., McBride, B.: RDF 1.1 Concepts and Abstract Syntax. W3C Recommendation (2014)

9. Gutierrez, C., Hurtado, C., Mendelzon, A.O.: Formal aspects of querying RDF databases. In: Proceedings of SWDB, pp. 293–307 (2003)

10. Harris, S., Seaborne, A., Prud'hommeaus, E.: SPARQL 1.1 Query Language. W3C Recommendation (2013)

11. Barbieri, D.F., Braga, D., Ceri, S., et al.: Querying RDF streams with C-SPARQL. ACM SIGMOD Rec. **39**(1), 20–26 (2010)

12. Le-Phuoc, D., Dao-Tran, M., Xavier Parreira, J., Hauswirth, M.: A native and adaptive approach for unified processing of linked streams and linked data. In: Aroyo, L., Welty, C., Alani, H., Taylor, J., Bernstein, A., Kagal, L., Noy, N., Blomqvist, E. (eds.) ISWC 2011. LNCS, vol. 7031, pp. 370–388. Springer, Heidelberg (2011). doi:10.1007/978-3-642-25073-6_24

13. Komazec, S., Cerri, D., Fensel, D.: Sparkwave: continuous schema-enhanced pattern matching over RDF data streams. In: ACM International Conference on Distributed Event-based Systems, pp. 58–68 (2012)

14. Allen, J.F.: Maintaining knowledge about temporal intervals. Readings Qual. Reason. Phys. Syst. **26**(11), 361–372 (1983)

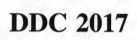

DDC 2017

Finding Optimal Team for Multi-skill Task in Spatial Crowdsourcing

Qian Tao[✉], Bowen Du, Tianshu Song, and Ke Xu

SKLSDE Lab and IRC, Beihang University, Beijing, China
{qiantao,dubowen,songts,kexu}@buaa.edu.cn

Abstract. These days, Online To Offline (O2O) platforms have been developing rapidly because of the popularization of smart phones and Mobile Internet. Spatial crowdsourcing, a burgeoning area in O2O market, is gaining more and more attention. It is a typical spatial crowdsourcing scenario in which an employer publishes a task and some workers will help him or her to accomplish it. However, most of previous work only considers the spatial information of workers and tasks, but ignores the individual variations among workers. In this paper, we raise a new problem called Software Development Team Formation (SDTF) problem, which aims to find a team of workers whose ability satisfies the requirement of the task. After showing the problem is NP-hard, we propose three greedy algorithms to approximately solve the problem. Besides, extensive experiments are conducted on synthetic and real datasets, which verify the effectiveness and efficiency of our algorithms.

Keywords: Spatial crowdsourcing · Task assignment · Team formation

1 Introduction

These days, with more and more people carrying mobile devices whenever and wherever, it is quite simple to acquire people's spatial and temporary information. As a result, many applications provide the services based on users' real-time spatial information and they are becoming more and more popular. These applications usually require some workers to help the employer to accomplish some tasks. For example, Didi Chuxing (http://www.didichuxing.com) arranges the drivers and provides users a convenient riding [12], Meituan (www.meituan.com) provides credible and fast food-delivering service, and etc. This area, called spatial crowdsourcing, is receiving plentiful attention.

Task assignment problem is one of the fundamental problems in the area of spatial crowdsourcing. Many studies on task assignment problem are published in recent years [5,11,13,14,16]. However, most of them only consider the spatial information of tasks and workers, but ignore the individual variations among workers. Different people may be good at/be weak in different tasks and tasks also contain certain requirements, which may be inadequate to some workers. Therefore, it is necessary to further consider individual variations of different

© Springer International Publishing AG 2017
S. Song et al. (Eds.): APWeb-WAIM 2017 Workshops, LNCS 10612, pp. 185–194, 2017.
https://doi.org/10.1007/978-3-319-69781-9_18

workers and special requirements of tasks. Same as [3], we use skills to represent the individual variations. Each worker is associated with a set of skills, which means that the worker is good at these skills. The task is also associated with a set of skills representing its special requirements.

In consideration of both workers and the employer, the objective of our work consists of two parts. On the one hand, workers need to move to the location of the task without any reward. On behalf of workers, we attempt to reduce the moving distance of workers. On the other hand, the employer tends to spend less money to accomplish the task. On behalf of the employer, we attempt to obtain a team with lower cost, on condition that the skills requirement is satisfied. As the problem definition in Sect. 3 shows, the objective function of our work considers not only the distance between the task and workers but also the cost.

Contributions. In summary, our contributions are:

- We propose a new problem called Software Development Team Formation (SDTF) problem and prove it to be NP-hard.
- Three greedy algorithms are provided to solve the SDTF problem.
- We verify the effectiveness and efficiency of the proposed algorithms through extensive experiments on synthetic and real datasets.

The rest of the paper is organized as follows. In Sect. 2, the problem is formally defined and proved to be NP-hard. In Sect. 3 three greedy algorithms are provided to solve the SDTF problem. Extensive experiments on real datasets are proposed in Sect. 4. Previous work related to our problem is presented in Sect. 5. Conclusions of our work are presented in Sect. 6.

2   Problem Statement

First, we introduce two basic concepts, task and coder. Then, we formally give the definition of the $\underline{S}$oftware $\underline{D}$evelopment $\underline{T}$eam $\underline{F}$ormation (SDTF) problem.

Definition 1 (Task). *A task t is defined as $<S, L>$, where $t.S$ is a set of skills which are indispensable to complete the software development task t, and $t.L$ is the location specified to meet up and talk about task t, which, for example, can be described by longitude and latitude.*

Similar to the definition of a task, a coder is formally defined as follows.

Definition 2 (Coder). *A coder c is defined as $<S, L, P>$, where $c.S$ is a set of skills mastered by coder c, $c.L$ is the location of coder c which is similar to that of a task t, and $c.P$ is the price of coder c.*

Briefly, a team of coders is feasible to a task, if the coders in the team can collaboratively accomplish the task.

Definition 3 (Feasible Team). *A team T is defined as a set of coders $\{c_1, c_2, ..., c_{|T|}\}$. T is a feasible team to a task t, if $\bigcup_{c_i \in T} c_i.S \supseteq t.S$.*

Example 1. Suppose that we have a task t about website development, where $t.S = \{\text{LINUX, DATABASE, CSS, HTML}\}$, and a universal of coders $C = \{c_1, c_2, c_3, c_4, c_5, c_6\}$. The skills set of every $c \in C$ is listed in Table 1. The team $T = \{c_2, c_4\}$ is a feasible team because $\bigcup_{c \in T} c.S = \{\text{CSS, LINUX, HTML,}$ DATABASE, JAVA$\}$, which is a superset of $t.S = \{\text{LINUX, DATABASE, CSS,}$ HTML$\}$.

We finally define our problem as follows.

Table 1. Coder profile

Coder	Skill	Price	Distance
c_1	{PYTHON, CSS, C++}	70	1000
c_2	{CSS, LINUX, HTML}	60	5000
c_3	{LINUX, HTML}	50	7000
c_4	{DATABASE, JAVA}	55	2000
c_5	{PYTHON, DATABASE, LINUX}	65	6000
c_6	{HTML, C#}	60	3000

Definition 4 (SDTF Problem). *Given a task t with a set of skills $t.S$ and a location $t.L$, and a universal of coders $C = \{c_1, c_2, ..., c_{|C|}\}$, each with a skills set $c_i.S$, a location $c_i.L$ and a price $c_i.P$ ($1 \leq i \leq |C|$), we want to find a team $T \subseteq C$ satisfying $\bigcup_{c \in T} c.S \supseteq t.S$, and we want to minimize $Cost = \alpha \cdot \max_{c \in T} |c.L, t.L| + (1 - \alpha) \cdot \sum_{c \in T} c.P$, where $|c.L, t.L|$ represents the distance between the location of coder c and task t, and $\alpha \in (0, 1)$ is a parameter to balance the weight between distance and price.*

Theorem 1. *The SDTF problem is NP-hard.*

Proof. We prove the theorem by the fact that a special case of SDTF problem can be reduced from the weighted set cover optimization problem. An instance of a weighted set cover problem consists of a set $U = \{1, 2, ..., n\}$, and a set $S = \{S_1, S_2, ..., S_{|U|}\}$, where $S_i \subseteq U$ for $1 \leq i \leq |U|$. Each S_i is associated with a positive value W_{S_i}, which can be viewed as the weight of S_i. The weighted set cover optimization problem is to find a subset S^* of S satisfying $\bigcup_{S_i \in S^*} S_i \supseteq U$ and $\sum_{S_i \in S^*} W_{S_i}$ is minimized. We consider a special case of SDTF problem, where the task and coders are located in the same position, and the skill set of the task is the universal of all skills. To reduce the weighted set cover optimization problem to the special case of SDTF problem, we observe that each element in U corresponds to a skill in $t.S$, each element in S_i corresponds to a skill in $c_i.S$, and the weight of S_i corresponds to the price of c_i. Since the task and all coders are at the same location, for every team T, $\max_{c \in T} |c.L, t.L|$ is 0, and we only need to minimize $\sum_{c \in T} c.P$. Obviously there exists a solution to the

---

Algorithm 1. PF-SDTF

---

input : Task t, Coders $C = \{c_1, c_2, ..., c_{|C|}\}$
output: A feasible team T

1   $T \leftarrow \emptyset$;
2 **while** T *is not feasible* **do**
3     $c \leftarrow \text{argmin}_{c \in C \,\&\&\, c.S \cap t.S \neq \emptyset} \frac{c.P}{|c.S \cap t.S|}$;
4     $T \leftarrow T \bigcup \{c\}$;
5     $t.S \leftarrow t.S - c.S$;
6 **return** T

---

weighted set cover optimization problem if and only if there exists a solution to the special case of SDTF problem, and we can obtain an instance of special case of SDTF problem from the instance of weighted set cover optimization problem in polynomial time. Therefore, the general case of SDTF problem is NP-hard.

3   Greedy Solutions for SDTF

In this section, we present three greedy algorithms to solve SDTF problem. The first two algorithms greedily choose a nearest/cheapest coder which can cover at least one uncovered skill. Because they only consider to partly optimize the object function, the solution is sometimes not good enough. Thus we propose the third greedy algorithm, which considers both price and distance when choosing a new coder.

3.1   PF-SDTF Greedy Algorithm

The idea of the first greedy algorithm PF-SDTF is to repeatedly add the coder with smallest price to the team while team is not feasible. The whole procedure of PF-SDTF is illustrated in Algorithm 1. We assume that there exists at least one feasible team. In line 1, we initialize an empty team T. In lines 2–5, when T is not feasible, we find a coder c who can cover at least one uncovered skill of task t and has the least value of $\frac{c.P}{|c.S \cap t.S|}$, add c to team T, and update $t.S$. Tie is broken by distance first, then arbitrarily. In line 6, we return the found feasible team T.

3.2   DistanceFirst-SDTF Greedy Algorithm

The idea of DF-SDTF is to repeatedly add the nearest coder to the team while the team is not feasible. The whole procedure of DF-SDTF is illustrated in Algorithm 2. We assume that there exists at least one feasible team. In line 1, we initialize an empty team T. In lines 2–5, when T is not feasible, we find a nearest coder c who can cover at least one uncovered skill of task t, add c to team T, and update $t.S$. Tie is broken by price first, then arbitrarily. In line 6, we return the found feasible team T.

Algorithm 2. DF-SDTF

---

input : $Task\ t, Coders\ C = \{c_1, c_2, ..., c_{|C|}\}$
output: A feasible team T
1 $T \leftarrow \emptyset$;
2 **while** T *is not feasible* **do**
3     $c \leftarrow \text{argmin}_{c \in C \&\& c.S \cap t.S \neq \emptyset} |c.L, t.L|$;
4     $T \leftarrow T \bigcup \{c\}$;
5     $t.S \leftarrow t.S - c.S$;
6 **return** T

Algorithm 3. DP-SDTF

---

input : $Task\ t, Coders\ C = \{c_1, c_2, ..., c_{|C|}\}$
output: A feasible team T
1 $T \leftarrow \emptyset$;
2 **while** T *is not feasible* **do**
3     $c \leftarrow \text{argmax}_{c \in C \&\& c.S \cap t.S \neq \emptyset} Utility(c, t, T)$;
4     $T \leftarrow T \bigcup \{c\}$;
5     $t.S \leftarrow t.S - c.S$;
6 **return** T

3.3 DP-SDTF Greedy Algorithm

The aforementioned two greedy algorithms are not effective, because they only try to optimize part of the object function. To optimize both distance and price at every step of iterations, we design a utility function $Utility$. Given a task t, current team T and a coder c, the definition of $Utility$ is

$$Utility(c, t, T) = \frac{|c.S \bigcap t.S|}{\alpha \cdot \Delta D(c, t, T) + (1 - \alpha) \cdot |c.P|} \tag{1}$$

where $\Delta D(c, t, T)$ represents the increment of maximum distance if c is added to team T, i.e., $\Delta D(c, t, T) = |c.L, t.L| - \max_{c' \in T} |c'.L, t.L|$ if $|c.L, t.L| > \max_{c' \in T} |c'.L, t.L|$, and $\Delta D(c, t, T) = 0$ if $|c.L, t.L| \leqslant \max_{c' \in T} |c'.L, t.L|$. As a matter of fact, the value of $\alpha \cdot \Delta D(c, t, T) + (1 - \alpha) \cdot |c.P|$ in Eq. (1) is the increment of the objective function. By using the utility function, we give the third greedy algorithm DP-SDTF. The whole procedure of DP-SDTF is illustrated in Algorithm 3. We assume that there exists at least one feasible team. In line 1, we initialize an empty team T. In lines 2–5, when T is not feasible, we find a coder c who gains the highest utility. Tie is broken by distance first, then arbitrarily. In line 6, we return the found feasible team T.

4 Evaluation

DataSet. We use real and synthetic datasets to evaluate our algorithms. Real data comes from CSTO (http://www.csto.com/). In CSTO dataset, each task is associated with a set of skills, and each coder is associated with a set of skills and an average price which can be deduced from the history data. We firstly analyze the price distribution using coders associated with price information. Figure 1 shows the relationship between the price and the number of skills mastered by the coder, where the horizontal axis represents skill number, and the vertical axis represents price. Except some expensive coder, we can observe than the price of a coder is uniform distributed between 0 to 5000, having nothing to do with the number of mastered skills. Since the CSTO data is not associated with location information, we generate the distance of each coder from the task following Uniform distribution. For synthetic data, based on the observation from real data set, we generate the price $c.P$ of coder c following uniform distribution. We assume each coder has 5 to 25 skills, which is common in reality. The distance from each coder to the task is generated following Uniform distribution. Statistics and configuration of synthetic data are illustrated in Table 2, where we mark our default settings in bold font.

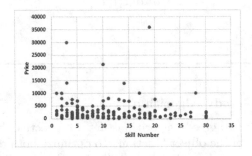

Fig. 1. Coder price distribution on CSTO.

Table 2. Synthetic dataset

Factor	Setting
α	0.1, 0.2, 0.3, 0.4, **0.5**, 0.6, 0.7, 0.8, 0.9
$\|t.Skills\|$	5, 6, 7, 8, 9, **10**, 11, 12, 13, 14, 15
$\|\bigcup_{c \in C} c.Skills\|$	100, 110, 120, 130, 140, **150**, 160, 170, 180, 190, 200
$\|C\|$	1W, 2W, 3W, 4W, **5W**, 6W, 7W, 8W, 9W, 10W

We evaluate our algorithms in terms of cost, running time and memory cost, and study the effect of varying parameters on the performance of the algorithms. The algorithms are implemented in C++, and the experiments were performed on a machine with Intel i7-4710 mq 2.50 GHZ 4-core CPU and 8 GB memory.

Effect of α. Figure 2a shows the effective of varying α. As α varies from 0.1 to 0.9, cost of DP-SDTF decreases smoothly, indicating $\sum_{c \in T} c.P$ contributes

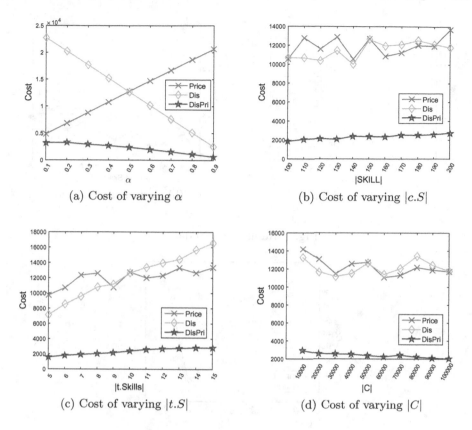

(a) Cost of varying α (b) Cost of varying $|c.S|$

(c) Cost of varying $|t.S|$ (d) Cost of varying $|C|$

Fig. 2. Results on synthetic data.

more compared with $\max_{c \in T} c.D$. Because DF-SDTF(PF-SDTF) algorithm only considers distance(price) when forming team T, when α is high(low), the perform is similar to DP-SDTF algorithm. However, when α decreases(increases), the performance of DF-SDTF(PF-SDTF) gets worse.

Effect of $|c.S|$, $|t.S|$ **and** $|C|$. We next show the effect of varying $|c.S|$, $|t.S|$ and $|C|$ in Fig. 2b, 2c and 2d. Because the default setting of α is 0.5, to find a good team, we need to consider distance and price simultaneously. We can observe that DP-SDTF algorithm performs best and costs of DF-SDTF and PF-SDTF are 3 to 4 times of that of DP-SDTF.

Real Data. Experiment results of real data are shown in Fig. 3. Figure 3a shows the effects of varying α, and Fig. 3b shows the effects of varying $|t.S|$. The effects of varying α are similar to the corresponding results of synthetic data. When varying $|t.S|$, costs of three algorithms on real data has some shaking, which is probably because the structure of CSTO dataset. As is showed in the figure, DP-SDTF algorithm still performs best.

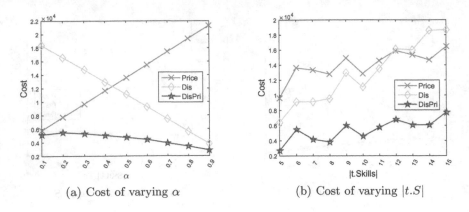

(a) Cost of varying α (b) Cost of varying $|t.S|$

Fig. 3. Results on real data.

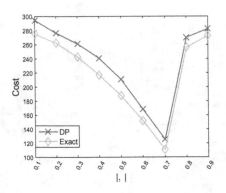

Fig. 4. DP-SDTF vs. exact results.

Compared with Exact Result. Because the SDTF problem is NP-hard, we only conduct experiments of small size of dataset to compare the output of our algorithm DP-SDTF with the exact solution. The setting is $t.S = 5$ and $|C| = 300$. The experiment result is shown in Fig. 4. We can observe that the performance of DP-SDTF is similar to the exact algorithm.

Conclusion. We conduct extensive experiments on both real and synthetic data. The results show that our algorithm can achieve similar performance with the exponential exact algorithm.

5   Related Work

The SDTF problem tackled in this paper contains knowledge of two domains: **Team Formation** and **Spatial Crowdsourcing**. On the one hand, The SDTF problem can be simplified to the task assignment problem if we ignore the skill constraint. On the other hand, it is exactly the most distinctive requirement that the skills of a team must cover the skills of the task. Related work of these two domains will be introduced in this section.

5.1 Team Formation

[7] firstly proposes the team formation problem. The problem requires a team of workers that: (1) its skills satisfy the requirement of the task; (2) the overall communication cost is minimum. In this paper, the NP-hardness of this problem is also proved. [8] extends the problem and associates each worker with a capacity, which represents the maximum number of assigned tasks to the worker. To solve the capacitated team formation problem, two approximation algorithms with proved guarantees are proposed. Different from [7,8] only including single task, [2] considers the team formation problem with multiple tasks and workers in both offline scenario and online scenario. While the above mentioned papers try to optimize the overall communication cost, [1] attempts to balance the workload among workers but merely treats the communication cost as a restrictive constraint. As the above shows, most of studies on team formation focus on skills satisfaction in communicative graph, while ignoring the influence of spatial information.

5.2 Spatial Crowdsoucing

The problem studied in this paper is the extension of task assignment problem in spatial crowdsourcing. It is called the server-assigned task assignment problem [5,6,11], in which workers are compulsory for the assigned tasks and cannot reject. [14] studies the online version of task assignment problem and provide an algorithm, which possesses constant competitive ratio under the random order model. Based on the original task assignment problem, [10,15] study the conflict-aware task assignment problem, in which tasks may conflict with each other and thus cannot be assigned to the same worker. Moreover, recent studies also consider to tackle the assignment problem and the planning problem at the same time. [9] attempts to solve the assignment problem and planning problem in the condition that each worker is associated with a travel budget.

Recently, [3,4] combine the task assignment problem and team formation problem, and propose a two-level-based framework to solve the problem. There are mainly two differences between [3,4] and our work: (1) there is no capacity constraint in our work, which means that there are more candidates in searching place; (2) the objective of our work considers both the distance between the task and workers and overall cost, while [3,4] only attempt to minimize the overall cost.

6   Conclusion

In this paper, we propose a novel crowdsourcing problem called Software Development Team Formation (SDTF) Problem. We prove that SDTF is NP-hard. Then, we design three greedy algorithms to solve the SDTF problem. The first two greedy algorithms, DF-SDTF and PF-SDTF, only consider part of the optimization object, and the performance is below the expectation. To overcome the shortcoming of the first two algorithms, we design the third greedy algorithm called DP-SDTF, which considers both part of the optimization goal comprehensively. We conduct extensive experiments to show the performance of our algorithms. The results show that our algorithm DP-SDTF achieves similar performance with exponential exact algorithm.

References

1. Anagnostopoulos, A., Becchetti, L., Castillo, C., Gionis, A., Leonardi, S.: Power in unity: forming teams in large-scale community systems. In: CIKM, pp. 599–608 (2010)
2. Anagnostopoulos, A., Becchetti, L., Castillo, C., Gionis, A., Leonardi, S.: Online team formation in social networks. In: WWW, pp. 839–848 (2012)
3. Gao, D., Tong, Y., She, J., Song, T., Chen, L., Xu, K.: Top-k team recommendation in spatial crowdsourcing. In: Cui, B., Zhang, N., Xu, J., Lian, X., Liu, D. (eds.) WAIM 2016. LNCS, vol. 9658, pp. 191–204. Springer, Cham (2016). doi:10.1007/978-3-319-39937-9_15
4. Gao, D., Tong, Y., She, J., Song, T., Chen, L., Xu, K.: Top-k team recommendation and its variants in spatial crowdsourcing. Data Sci. Eng. **2**(2), 136–150 (2017)
5. Kazemi, L., Shahabi, C.: Geocrowd: enabling query answering with spatial crowdsourcing. In: GIS, pp. 189–198 (2012)
6. Kazemi, L., Shahabi, C., Chen, L.: Geotrucrowd: trustworthy query answering with spatial crowdsourcing. In: GIS, pp. 304–313 (2013)
7. Lappas, T., Liu, K., Terzi, E.: Finding a team of experts in social networks. In: SIGKDD, pp. 467–476 (2009)
8. Majumder, A., Datta, S., Naidu, K.: Capacitated team formation problem on social networks. In: SIGKDD, pp. 1005–1013 (2012)
9. She, J., Tong, Y., Chen, L.: Utility-aware social event-participant planning. In: SIGMOD, pp. 1629–1643 (2015)
10. She, J., Tong, Y., Chen, L., Cao, C.C.: Conflict-aware event-participant arrangement and its variant for online setting. IEEE Trans. Knowl. Data Eng. **28**(9), 2281–2295 (2016)
11. To, H., Shahabi, C., Kazemi, L.: A server-assigned spatial crowdsourcing framework. ACM Trans. Spat. Algorithms Syst. **1**(1), 2 (2015)
12. Tong, Y., Chen, Y., Zhou, Z., Chen, L., Wang, J., Yang, Q., Ye, J., Lv, W.: The simpler the better: a unified approach to predicting original taxi demands based on large-scale online platforms. In: SIGKDD, pp. 1653–1662 (2017)
13. Tong, Y., She, J., Ding, B., Chen, L., Wo, T., Xu, K.: Online minimum matching in real-time spatial data: experiments and analysis. Proc. VLDB Endow. **9**(12), 1053–1064 (2016)
14. Tong, Y., She, J., Ding, B., Wang, L., Chen, L.: Online mobile micro-task allocation in spatial crowdsourcing. In: ICDE, pp. 49–60 (2016)
15. Tong, Y., She, J., Meng, R.: Bottleneck-aware arrangement over event-based social networks: the max-min approach. World Wide Web: Internet Web Inf. Syst. **19**(6), 1151–1177 (2016)
16. Tong, Y., Wang, L., Zhou, Z., Ding, B., Chen, L., Ye, J., Xu, K.: Flexible online task assignment in real-time spatial data. Proc. VLDB Endow. **10**(11), 1334–1345 (2017)

CI-Bot: A Hybrid Chatbot Enhanced by Crowdsourcing

Xulei Liang, Rong Ding, Mengxiang Lin$^{(\boxtimes)}$, Lei Li, Xingchi Li, and Song Lu

State Key Laboratory of Software Development Environment, Beihang University,
Beijing 100191, China
linmx@buaa.edu.cn

Abstract. Question and answer website is an effective way for people to get information from others. Recently, chatbot has been more and more widely used. In this paper, we propose CI-Bot, a Crowd-Intelligence-chatBot. CI-Bot is a hybrid intelligent chatbot, in which crowdsourcing is introduced on the basis of a chatbot. When receiving a problem, the conversational partner of CI-Bot first tries to solve it automatically. If the question is beyond the knowledge of CI-Bot, expert recommender would find out experts it knows and consults them. Ultimately, the problem would be solved and the answers generated by the experts are added to a corpus, to increase the ability of CI-Bot. We implemented a prototype on the top of Hubot and Wechat. The preliminary experiment results validate the effectiveness of CI-Bot.

1 Introduction

Asking question is an effective way for people to learn from each other. When facing with a problem, we may consult our friends or post the question to a Question and answer (Q&A) site to get help from persons on the Internet. Besides, people could use search engines to search for related information or even ask an intelligent chatbot to get the answer immediately.

There are advantages and disadvantages of these two approaches. By voting, bidding or other incentive mechanisms, Q&A sites provide high-quality contents via crowdsourcing [1] although it may take a few hours or even days. Meanwhile, chatbots are easy to interact with and able to make timely responses. Chatbots designed to conduct conversations to make people feel comfortable are hard to solve and explain professional problems.

When dealing with problems, everyone wants to obtain timely help, regardless of whether it comes from human or machine. Based on the above idea, we propose CI-Bot, a Crowd-Intelligence-chatBot. The basic idea behind CI-Bot is: "Something though I don't know, I know who most likely to know." When receiving a problem, CI-Bot first tries to solve it automatically. If it fails, CI-Bot would find out experts it knows and consult them. The answers given by the experts are integrated into a final answer and the final answer is added to a corpus to improve the ability of CI-Bot.

The main contributions of this paper are as follows:

© Springer International Publishing AG 2017
S. Song et al. (Eds.): APWeb-WAIM 2017 Workshops, LNCS 10612, pp. 195–203, 2017.
https://doi.org/10.1007/978-3-319-69781-9_19

1. A framework of CI-Bot, a hybrid intelligent chatbot system;
2. A prototype is implemented and deployed based on Wechat;
3. Experiments are conducted to evaluate the effectiveness of CI-Bot.

The remainder of this paper is organized as follows. Section 2 introduces the framework of CI-Bot and discusses the essential parts of it. Implementation and deployment of the prototype are presented in Sect. 3, which is followed in Sect. 4 by a preliminary experiment and a discussion of the potential applications of CI-Bot. Surveys of related work are given in Sect. 5. Section 6 presents the conclusion.

2   CI-Bot Framework

In this section, we first introduce the structure and workflow of CI-Bot and then discuss three essential parts in it: Conversational Partner, Expert Recommender and Answer Integration.

2.1   Structure and Workflow

As shown in Fig. 1, CI-Bot consists of AI module and CI module in a high level. The AI module of CI-Bot has mainly two functions: chatting with users and answering questions based on corpus. The corpus is a set of dialog pairs, and the Conversational Partner receives the message from users and generates the response according to the corpus. Two different techniques are applied in the Conversational Partner: the information retrieval method and the neural network method.

The ability of AI module is relatively limited. For questions AI module cannot answer properly, CI-Bot would use the CI module to get solutions from people. The CI module includes three parts: Crowd, Expert Recommender and Answer Integration. Crowd is the user group of CI-Bot. Every user in the Crowd could ask or answer a question. When AI module sends a question to the CI module, the Expert Recommender would check information of every user and find out users who have knowledge of relevant areas and forward the question to them. After received answers from the experts, the Answer Integration combines or compares the answers to conclude a final answer. The final answer would be sent to the asker and added to the corpus in AI module.

The workflow of CI-Bot is as follows:

1. Receive Message: A user posts a message to CI-Bot by a chatting application.
2. Chat Reply: If the message is a chat message, neural network method is used to produce the response.
3. Machine Q&A: If the message is a question, retrieval-based method is used to generate a proper answer. If the user is satisfied with the answer, the conversation stops. If the user refused the answer, or CI-Bot fails to generate an answer, proceed to the next step.
4. Expert Recommend: Find related experts and send the question to them.

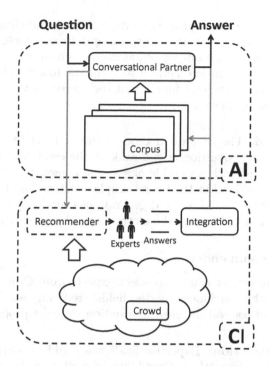

Fig. 1. The framework of CI-Bot

5. Integrate Answers: Collect answers from experts and integrate the answers in to a final answer. Send the final answer to the asker.
6. Motivate Experts: If the final answer is accepted by the asker, reward corresponding experts.
7. Corpus Updating: Add the question and the accepted answer to the corpus.

2.2 Conversational Partner

The Conversational Partner is devoted to response a message automatically. It supports two kinds of dialog: chat and question answering. After receiving a message, CI-Bot first distinguishes what kind of message it is. If it is a chat message, CI-Bot inputs the message to a trained neural network to produce a response. If it is a question, the retrieval method will search for answers from similar questions.

Neural Network Method. The main idea of neural network method [2] is to map the input sequence into a fixed dimension vector by using Long Short Term Memory(LSTM) layer, then use another LSTM layer to extract the output sequence from the vector. The actual model is implemented by multi layer LSTM neural network. The model reads the input sequence one token at a time, then the input layer will map this sequence to a vector with certain length. The output layer predicts the output sequence one token at a time as well. In the training process, since the input and output sequences are given at the same time, we could use the back propagation method to train the model. The final model is obtained by minimizing the cost function of the correct output sequence in the case of a given input sequence.

Retrieval Method. The retrieval method [3] stores the Q&A pairs in a corpus. When the users input a question, it will rank all the candidate responses with a valuation function and pick a suitable one as the answer to return. One of the major strengths of retrieval method is its simple construction. It is easy to build a retrieval model with the corpus and generate answers with details. If there is no similar question in the corpus, the method would not return an answer.

2.3   Expert Recommender

The Expert Recommender is used to select experts from Crowd. CI-Bot stores user information, which includes expertise fields, credibility and motivation. The recommend method uses all the user information to select proper experts.

Information Extraction. Expertise fields are could be obtained from user selection or text mining method. Credibility is similar to the prestige value. It is inferred from the user's income and received answers. Motivation of a user is inferred from the number of questions user answered and the response time after he/she viewed the question.

Recommend Method. CI-Bot selects top N most relevant users as experts. There are three main steps in this process. In the first step, CI-Bot calculates the similarity between expertise fields of users and the content of the question, and ranks all the similarities. In the second step, CI-Bot ranks the users according to the credibility and motivation of users so as to choose the experts who are more willing to answer the question. Finally, CI-Bot calculates integrated ranking and selects the top N users as experts.

2.4   Answer Integration

To ensure the quality of the answers, CI-Bot generates the final answer from multiple replies and sends it back to the questioner.

Noise Filtering. Since the answers are submitted by various experts, there must be some inevitable noises in responses, such as malicious words or meaningless answers. Malicious words are filtered by searching key words in malicious vocabularies. For meaningless answers, CI-Bot calculates the semantic relevance between questions and answers. If they have rather low relevance, CI-Bot would assume that the answer is invalid to this question.

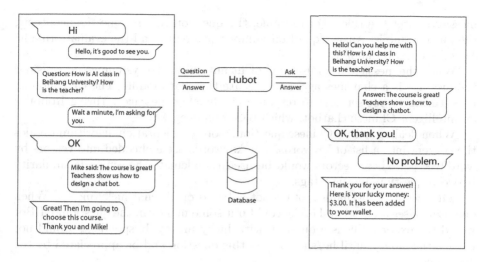

Fig. 2. The deployment of prototype

Integrate Answers. There are various integration methods for different kinds of questions. For the selective questions, CI-Bot counts the number of each option. The proportion of different options is shown to the user for reference. Numerical questions, such as estimating the weight of a cow, can be solved by averaging or taking the mode of the answers. As for explanatory questions, the best answer is chose according to several factors, such as the length of the answer and the response time.

3   Implementation and Deployment

We implemented a prototype of CI-Bot based on Hubot[1], an popular open-sourced chatbot framework. Wechat[2], the most widely used social platform in China, is used as our interaction platform. CI-Bot is connected with Wechat through a Wechat adapter[3] and an account is registered specially for the application. Since Hubot does not have the function to generate conversational content, we developed several modules in Hubot to response the user behaviors. The prototype does not need servers, therefore a PC is used to deploy the application. Furthermore, we use MySQL[4] to store all the related data, including the corpus, the user information and the status of each question. Figure 2 shows our prototype deployment.

In the current implementation, CI-Bot checks user input by a set of rules. Users are welcomed to raise any questions in natural languages as long as their

---

[1] Hubot: https://hubot.github.com/.
[2] Wechat: https://weixin.qq.com/.
[3] Wechat-adapter: https://github.com/KasperDeng/Hubot-WeChat.
[4] MySQL: https://www.mysql.com/.

messages meet the rules. For example, the question rule is: Question:[content], the answer rule is: Answer[question number]:[content], and the evaluate rule is: Yes/No [question number].

When the message from user do not fit any rules, the system would consider the message is a chat message, and execute the chat module. The chat module uses Turing Robot service[5] to response all the chat messages. Turing Robot is an intelligent Chinese chatbot, which based on deepQA.

When dealing with a Chinese question, word segmentation algorithm divides the content into a list of keywords. The keywords are embedded into vectors by word2vec [4]. These vectors would be used to calculate the semantic similarity between question and user tags.

The incentive mechanism of CI-Bot is based on Wechat lucky money. When one user raises a question, he/she could put some money in the question. CI-Bot would tell experts this is a question with lucky money. Respondents would not receive the money until he/she answers this question and be appreciated by the questioner.

The dialog boxes in Fig. 2 shows a relatively complete conversation of the prototype. Just using a dialog interface, user could execute all the actions. In order to make it easier to understand, Fig. 2 omits the details of answer integration and question number.

4 Experiment and Discussion

4.1 Experiments

We recruited a number of volunteers to evaluate the effectiveness of our prototype. Volunteers are allowed to raise questions with lucky money in Wechat. Each of them had 10.00 yuan as initial assets which could only be used in questioning.

At the beginning of the experiments, there is no Q&A pairs in the corpus. So the prototype only responses to chat messages or consults people. As the experiments went on, more and more Q&A pairs are added into the corpus, and the ability of CI-Bot have increased. There are 43 participants in the experiment, excluding authors. Totally 736 pieces of conversations, 40 times of questioning and 49 answers are recorded. All the accepted answers are added to the corpus of CI-Bot, and become the knowledge of CI-Bot. If someone ask the same question, CI-Bot could answer questions without consulting people.

Three quarters of the problems are answered in this experiment. Figure 3 shows the response time of the first answers. 40% questions are answered within 50 s. The average time taken to receive the first answer is 5 min.

Figure 4 shows the length of the answers. Four fifths answers have less than 40 words, showing that length of answers may be restricted due to the chatting mechanism. There are 14 of 16 first returned answers are accepted by askers, while only 2 of them are rejected. Rate of satisfactory are 35% in terms of all 40 questions. It is worth to note that the subjects of questions involve different domains such as movie, computer, music and management.

---

[5] Turing Robot: http://www.tuling123.com/.

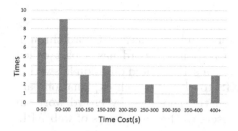
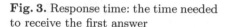
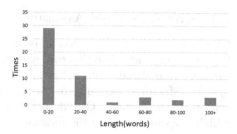

Fig. 3. Response time: the time needed to receive the first answer

Fig. 4. Answer length: the number of words in answers.

4.2 Discussion

At the end of the experiments, we make a simple survey of the usage and the application scenarios of CI-Bot. Most volunteers think it is convenient to interact with CI-Bot.

Many students offer us a lot idea about the potential applications. One of the promising applications is the image labelling. As the development of machine learning and computer vision, image lagelling has been in great demand. CI-Bot may serve as a platform for people to annotate images conveniently. Another potential application is to transfer information timely on a large scale such as employment information or community news.

Of course, the volunteers also offer some suggestions of the prototype. These suggestions point out our future work such as the knowledge management of CI-Bot. Our next step is to build knowledge graph in our system, Which is the structured information of professional knowledge.

5 Related Work

5.1 Crowd-Powered Search Engine

Since it is hard for computers to process unstructured information, some systems retrieval information through crowdsourcing. By hiring workers on Amazon Mechanical Turk (AMT), Crowd DB [5] tries to search unspecified sentences. KURATOR [6] is used to arrange family photos, videos, etc. KURATOR shows that crowd workers could deal problems with personal inclination. CRQA [7] is also a hybrid intellectual Q&A system. Candidate answers from the Internet are ranked by AMT workers. The research of CRQA proved that in a limited time, crowdsourcing could make high quality votes. The work of [8] answer questions raised in social network through the crowd of volunteers. The answer of volunteers are almost as good as that of friends.

5.2   Crowd-Powered Assistant

Chord [9] is a crowd-powered assistant. Users send messages to Chord through Google Now and Chord recruit workers to response. Chord has a part called Memory to maintain the context. AXIS [10] is a self-adopted question answering system. Machine learning are used in AXIS to choose which explanations are sent to later learners. Guardian [11] is a chatbot powered by crowd work, in which workers translate the message from users to the parameters of web API.

5.3   Real-Time Crowdsourcing

With the development of smartphones, recent studies of real-time crowdsourcing first focus on the real-time task assignment problem [12–14]. In addition, VizWiz [15] system helps the blind to acquire information about the environment. Volunteers offer their help by real-time video on smart phone. Vizwiz has ever helped 5000 blinds. Similarly, blinds could call for help using BeMyEyes app. Scribe [16] records live speech with the help of crowd working and machine learning. Scribe could response within 4 s.

6   Conclusion

This paper proposed CI-Bot, a Crowd-Intelligence-chatBot. CI-Bot could benefit from the artificial intelligence and the crowd intelligence. A prototype is implemented and deployed. The preliminary experiment results are promising in terms of response time and conversational scope.

References

1. Li, W., Wu, W., Wang, H., Cheng, X., Chen, H., Zhou, Z., Ding, R.: Crowd intelligence in AI 2.0 era. Front. Inf. Technol. Electron. Eng. **18**, 15–43 (2017)
2. Sutskever, I., Vinyals, O., Le, Q.V.: Sequence to sequence learning with neural networks. In: Advances in Neural Information Processing Systems, pp. 3104–3112 (2014)
3. Ji, Z., Lu, Z., Li, H.: An information retrieval approach to short text conversation. arXiv preprint. arXiv:1408.6988 (2014)
4. Goldberg, Y., Levy, O.: word2vec explained: deriving Mikolov et al.'s negative-sampling word-embedding method. arXiv preprint. arXiv:1402.3722 (2014)
5. Franklin, M.J., Kossmann, D., Kraska, T., Ramesh, S., Xin, R.: CrowdDB: answering queries with crowdsourcing. In: Proceedings of the 2011 ACM SIGMOD International Conference on Management of Data. ACM, pp. 61–72 (2011)
6. Merritt, D., Jones, J., Ackerman, M.S., Lasecki, W.S.: Kurator: using the crowd to help families with personal curation tasks (2017)
7. Savenkov, D., Weitzner, S., Agichtein, E.: Crowdsourcing for (almost) real-time question answering. In: Workshop on Human-Computer Question Answering, NAACL (2016)
8. Jeong, J.W., Morris, M.R., Teevan, J., Liebling, D.J.: A crowd-powered socially embedded search engine. In: ICWSM (2013)

9. Huang, T.H.K., Lasecki, W.S., Azaria, A., Bigham, J.P.: "Is there anything else I can help you with?" challenges in deploying an on-demand crowd-powered conversational agent. In: Fourth AAAI Conference on Human Computation and Crowdsourcing (2016)

10. Williams, J.J., Kim, J., Rafferty, A., Maldonado, S., Gajos, K.Z., Lasecki, W.S., Heffernan, N.: Axis: Generating explanations at scale with learnersourcing and machine learning. In: Proceedings of the Third ACM Conference on Learning@ Scale. ACM pp. 379–388 (2016)

11. Huang, T.H.K., Lasecki, W.S., Bigham, J.P.: Guardian: a crowd-powered spoken dialog system for web APIs. In: Third AAAI conference on human computation and crowdsourcing (2015)

12. Tong, Y., She, J., Ding, B., Chen, L., Wo, T., Xu, K.: Online minimum matching in real-time spatial data: experiments and analysis. Proc. VLDB Endow. **9**, 1053–1064 (2016)

13. Tong, Y., She, J., Ding, B., Wang, L., Chen, L.: Online mobile micro-task allocation in spatial crowdsourcing. In: Proceedings of the 32nd International Conference on Data Engineering, pp. 49–60 (2016)

14. Tong, Y., Wang, L., Zhou, Z., Ding, B., Chen, L., Ye, J., Xu, K.: Flexible online task assignment in real-time spatial data. Proc. VLDB Endow. **10**, 1334–1345 (2017)

15. Bigham, J.P., Jayant, C., Ji, H., Little, G., Miller, A., Miller, R.C., Miller, R., Tatarowicz, A., White, B., White, S., et al.: Vizwiz: nearly real-time answers to visual questions. In: Proceedings of the 23nd Annual ACM Symposium on User Interface Software and Technology. ACM, pp. 333–342 (2010)

16. Lasecki, W.S., Miller, C.D., Naim, I., Kushalnagar, R., Sadilek, A., Gildea, D., Bigham, J.P.: Scribe: deep integration of human and machine intelligence to caption speech in real time. Commun. ACM **60** (2017)

17. Yang, L., Qiu, M., Gottipati, S., Zhu, F., Jiang, J., Sun, H., Chen, Z.: CQArank: jointly model topics and expertise in community question answering. In: Proceedings of the 22nd ACM International Conference on Information and Knowledge Management. ACM, pp. 99–108 (2013)

18. Kittur, A., Nickerson, J.V., Bernstein, M., Gerber, E., Shaw, A., Zimmerman, J., Lease, M., Horton, J.: The future of crowd work. In: Proceedings of the 2013 Conference on Computer Supported Cooperative Work. ACM pp. 1301–1318 (2013)

Group Formation Based on Crowdsourced Top-k Recommendation

Yunpeng Gao[1], Wei Cai[2(✉)], and Kuiyang Liang[1]

[1] State Key Laboratory of Software Development Environment,
Beihang University, Beijing, China
[2] Beijing Science and Technology Information Center, Beijing, China
caiw@bsw.gov.cn

Abstract. There has been significant recent interest in the area of group recommendations, where, given groups of users of a recommender system, one wants to recommend $top - k$ items to a group to achieve an object such as minimizing the maximum disagreement between group members, according to a chosen semantics of group satisfaction. We consider the complementary problem of how to form groups such that the users in the formed groups are as even satisfied with the suggested $top - k$ recommendations as possible. Thanks to emerging crowdsourcing platforms, e.g., Amazon Mechanical Turk and CrowdFlower, we have easy access to workers' $top - N$ recommendation. Here dealing with the ranking data is a big challenge, as quite few methods to solve this issue. We assume that the recommendations will be generated according to minimize the maximum disagreement between group users utilizing the kendall distance. Rather than assuming groups are given, or rely on ad hoc group formation dynamics, our framework allows a strategic approach for forming groups of users in order to minimize the maximum disagreement. Furthermore, we develop efficient algorithms for group formation under the minmax object. We validate our results and demonstrate the scalability and effectiveness of our group formation algorithms on both real and synthetic data sets.

1 Introduction

Crowdsourcing is a specific sourcing model in which individuals or organizations use contributions from Internet users to obtain needed services or ideas. Crowdsourcing in the form of idea competitions or innovation contests provides a way for organizations to learn beyond what their "base of minds" of employees provides [1,2].

There is a proliferation of group recommender systems that cope with the challenge of addressing recommendations for groups of users. FlyTrap [3] recommends music to be played in a public room. What all these recommender systems have in common is that they assume that the groups are ad hoc, are formed organically and are provided as inputs, and focus on designing the most appropriate recommendation semantics and effective algorithms. In all these systems,

© Springer International Publishing AG 2017
S. Song et al. (Eds.): APWeb-WAIM 2017 Workshops, LNCS 10612, pp. 204–213, 2017.
https://doi.org/10.1007/978-3-319-69781-9_20

all members of a group are recommended a common list of items to consume. Indeed, designing semantics to recommend items to ad-hoc groups has been a subject of recent research [4–7], and several algorithms have been developed for recommending items personalized to given groups of users [6,8].

On the other hand, we study the flip problem and address the question, if group recommender systems follow one of these existing popular semantics, how best can we form groups to minimize the maximum disagreement between users. In fact, we pose this as an optimization problem to form groups in a principled manner, such that, after the groups are formed and the users inside the group are recommended an item set to consume together, following the Kemeny ranking aggregating method and group recommendation algorithms, they are as even satisfied as possible.

Our strategic group formation is potentially of interest to all group recommender system applications, as long as they use certain recommendation semantics. Instead of ad-hoc group formation [3,6,9], or grouping individuals based on similarity in preferences [10], or meta-data [11–13] , we explicitly embed the underlying group recommendation semantics in the group formation phase, which may dramatically improve user satisfaction. Our focus in this work is to formalize how to form a set of user groups for a given population, such that, the the maximum disagreement of all groups w.r.t. their recommended $top - k$ item lists, generated according to group recommendation semantics, is minimized.

Our work is also orthogonal to existing market based strategies [14] on daily deals sites, such as Groupon and LivingSocial. Most of these works focus on recommending deals to users. They rely on incentivizing formation of groups via price discounting. The utility of such recommendation strategies is to maximize revenue, whereas, our group formation problem is purely designed to minimize users maximum disagreement. We elaborate on this interesting but orthogonal research direction further in the related work section.

In summary, we make the following contributions:

- In crowdsourcing setting, workers are required to recommend N items in a whole itemset, (ranking orders), which are used to group workers.
- We initiate the study of how to form groups in the context of recommender systems, considering popular group recommendation semantics. We formalize the task as a minmax problem, with the objective to form groups, such that the maximum group disagreement w.r.t. the suggested group recommendation is minimized.
- Here we use kendall distance to define disagreement between users' preference ranking, which is a solution to deal with the tricky problem. Then we propose group distance and the group formation which formalize our object.
- We conduct a comprehensive experimental study on both real world data sets and synthetic data sets. Our proposed clustering algorithm outperform to others, which is efficient and small group formation value, leading to relatively balanced group sizes, as even satisfied as possible.

2   Judgment Problem

In a crowdsourcing platform, given an item set $I = \{i_1, i_2, \ldots, i_m\}$ and a worker set $U = \{u_1, u_2, \ldots, u_n\}$, workers are required to recommend N items in a whole itemset. Each workers preference on I is $\pi_N = \{i_1, i_2, \ldots, i_N\}, N \ll m, p \in U$, which means for user u_p, item i_s is preferred to item i_t, $s < t$. We solve the problem of separating user set U to groups to make the maximized group disagreement is minimum.

Ranking distance. The ranking distance between two users ranking π_p, π_o is defined as follows:

$$D_{\pi_{p,o}} = \sum_{i=1}^{N} \sum_{j=1}^{N} 1\{\pi_p(i) < \pi_p(j), \pi_o(i) > \pi_o(j)\} \tag{1}$$

where $\pi_p(i)$ is the rank of item i in users preference ranking π, $D_{\pi_{p,o}}$ is the disagreement between the two rankings.

Group distance. Given user group $g = \{u_1, u_2, \ldots, u_s\}$, $s \leq m$ and the users ranking $\pi = (\pi_1, \pi_2, \ldots, \pi_s\}$, the group distance is defined as the maximized ranking distance between two users in the group, which is defined as follows:

$$GD_g = \max_{\pi_i, \pi_j \in \pi} D_{\pi_i, \pi_j} \tag{2}$$

Group Formation. Given an item set $I = \{i_1, i_2, \ldots, i_m\}$ and a user set $U = \{u_1, u_2, \ldots, u_n\}$, each users preference on I is $\pi_N = \{i_1, i_2, \ldots, i_N\}$, $N \ll m$, $p \in U$. To create a set of l non-overlapping groups, which each group g is associate with a $top-k$ item set I_g^k in accordance with group distance, the group formation is created when the group distance is minimized, which is define as follows:

$$GF = \min_{i \leq l} D_{g_i} \tag{3}$$

3   Solutions

3.1   Random Method

Borda partially ranking aggregation method is used to get π^* and the $top-k$ item set. Then randomly put the users in separate groups, so the objective GF and the groups are obtained.

Performance improvements: The performance of random method can be improved by processing the score computations with local search technique. We call it RandomLS. For sake of simplicity, we omit the implementation details.

Algorithm 1. Random method

Input: $I = \{i_1, i_2, \ldots, i_m\}, U = \{u_1, u_2, \ldots, u_n\}, \pi_p = \{i_1, i_2, \ldots, i_q\}$

Output: group $G = \{g_1, g_2, \ldots, g_l\}$

1: $\pi^* = Borda(\pi)$
2: $I_g^k = \{\pi^*(1), \pi^*(2), \ldots, \pi^*(k)\}$
3: $G_i = randomGroup(U, l)$ // random separate the user set U to l groups.
4: **for** each $G_i = \{g_1', g_2', \ldots, g_l'\}$ **do**
5:    **if** $GF_i = \min\limits_{i \leq l} D_{g_i}$ **then**
6:       $G = G_i$
7:    **end if**
8: **end for**
9: Return G

Algorithm 2. Greedy method

Input: $I = \{i_1, i_2, \ldots, i_m\}, U = \{u_1, u_2, \ldots, u_n\}, \pi_p = \{i_1, i_2, \ldots, i_q\}$

Output: group $G = \{g_1, g_2, \ldots, g_l\}$

1: Begin
2: $\pi^* = Borda(\pi)$
3: $I_g^k = \{\pi^*(1), \pi^*(2), \ldots, \pi^*(k)\}$
4: $\tau_1 = \min D_{\pi, \pi^*}$
5: $g_1 = \pi_i$
6: $\pi' = \pi - g_1$
7: **for** each j **do**
8:    $\tau_j = \min D_{\pi', \pi^*}$
9:    **if** $D_{\tau_j, \pi'} \leq \tau_1$ **then**
10:       $g_1 = g_1 \cup \tau_j$
11:    **end if**
12: **end for**
13: Return G

3.2   Greedy Method

Here π^* and the $top - k$ item set are obtained like the above methods. According to the ranking distance between π and π^*, rearrange the π_i in descending order. In order to get smaller group formation, we iteratively group the users using small ranking distance, and finally derive the group result.

3.3   Clustering Method

Here the clustering method is used to cluster the users in a group by their preference similarity. Here we employ the $k - means$ clustering algorithm.

4   Data Sets

In this section, we discuss the data sets that we used for our experiments. In collecting these data sets, we aimed for a variety of lengths (n), number of

---

Algorithm 3. Clustering method

    Input: $I = \{i_1, i_2, \ldots, i_m\}, U = \{u_1, u_2, \ldots, u_n\}, \pi_p = \{i_1, i_2, \ldots, i_q\}$
    Output: group $G = \{g_1, g_2, \ldots, g_l\}$
1: Begin
2: $\pi^* = Borda(\pi)$
3: $I_g^k = \{\pi^*(1), \pi^*(2), \ldots, \pi^*(k)\}$
4: $G = cluster(U, l)$ // k-means clustering algorithm
5: $GF = \min(C_1, C_2, \ldots, C_l)$ // obtain the minimum cluster
6: Return G

---

rankings (N), and degrees of consensus. To this end, we used real world and synthetic data sets. We first describe the real-world and then move on to the synthetic data sets.

4.1 Real World Data Sets

We procured several real world data sets, aiming for problems where a Kemeny ranking would be meaningful, but challenging to find.

Last.fm Data Set. In order to, verify the quality of the recommendation semantics, our algorithm was tested using 1 K Last.fm data set, which is widely used to evaluate the recommendation algorithm. This data set represents the full listening history (till May, 5th 2009) for nearly 1,000 users. We obtained the data from http://labrosa.ee.columbia.edu/millionsong/lastfm.

4.2 Synthetic Data Sets

We generated the following data sets in order to more thoroughly investigate algorithm performance on data sets of varying length, number of rankings, and consensus.

Mallows' Model Data Sets. We drew samples from a Mallows' model (MM) with various parameter settings. The Mallows' model is an exponential model over rankings introduced by Mallows [15] and is given by:

$$P_{\pi_0, \theta}(\pi) = \frac{exp(-\theta d(\pi, \pi_0))}{Z}$$

$$Z = \prod_{i=1}^{n-1} \frac{1 - exp((-n - i + 1)\theta_i)}{1 - exp(-\theta_i)} \tag{4}$$

In the above, π_0 is the central ranking of the model and $\theta \geq 0$. The parameter θ of the Mallows' model quantifies the concentration of the distribution around its peak π_0. For $\theta = 0$ the distribution is uniform (no peak, and no consensus), and for larger θ the distribution becomes increasingly peaked. Therefore, we expect that the larger the θ value, the stronger the consensus in the data, and therefore the easier the Kemeny ranking problem. We shall see that our experiments support this intuition. Also, from the point of view of an algorithm, increasing the number of items n will make the problem harder. The values for n, θ, N that we used are listed in Table 1. In selecting these parameters, we aimed to have a range of difficulties, with more emphasis on the challenging and hard problems.

Plackett-Luce model data sets. We also drew samples from a Plackett-Luce (P-L) model [16]. The P-L model is a probability distribution over rankings which is given by:

$$P_s(\pi) = \prod_{i=1}^{n} \frac{S_{\pi^{-1}(i)}}{Z_i}$$

$$Z_i = \sum_{i \leq j}^{n} S_{\pi^{-1}(j)}$$

(5)

where $S_i > 0$ are parameters, and $\pi^{-1}(i)$ returns the alternative in rank i of the permutations. Table 1 list the parameters used to generate the Mallows model and Plackett-Luce data sets.

Table 1. The parameters used to generate the Mallows model and Plackett-Luce datasets.

N	n	$\theta(MM)$	$S_{\pi^{-1}(i)}(P-L)$
100	10	0.001	10,9,...,2,1
100	50	0.001	50,49,...,2,1
100	10	0.01	—
100	50	0.01	—
100	10	0.1	—
100	50	0.1	—
5000	10	0.1	—
5000	50	0.1	—

5 Experiments

5.1 Result on Last.fm Data Set

We ran each algorithm described above on each data set and recorded the running time, result GF, and other variables. Because the optimal method is too time-consuming for all the data sets, we just compare the other four methods(Random, RandomLS, Greedy, Cluster) (Fig. 1).

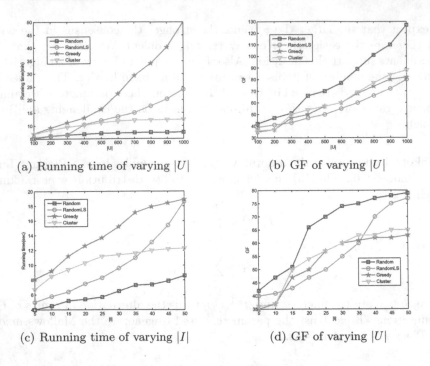

(a) Running time of varying $|U|$ (b) GF of varying $|U|$

(c) Running time of varying $|I|$ (d) GF of varying $|U|$

Fig. 1. Last.fm data set result on varying $|U|$ and $|I|$

As in last.fm data set, figurelast.fm(a) shows that Random and RandomLS methods are better other two methods regard running time of varying $|U|$. Particularly, We can see that the running time of random method nearly increases no time when the number of users changes. The Cluster method performs better than greedy method, which gets a sharply increase as the users number goes up. From figure last.fm(b), we can see that greedy and clustering methods are much better than random method, which means that from group formation perspective, greedy and clustering methods have a good performance. This may happen because the greedy method iteratively chooses the minimized ranking distance between two rankings, which is very time-consuming; however, it towards to relatively better group formation. The running time and GF of the four algorithms change less on varying $|I|$ than $|U|$, which indicates that the number of users may have greater impact on the group formation problem.

5.2 Result on Mallows' Model Data Sets and P-L Data Sets

To test our proposed algorithms scalability, we obtain the result on Mallows' model data sets and P-L data sets.

Except Greedy algorithm runs longer, the other three algorithms have a good performance even the number of users go to 5000. On the other hand, the GF value changes not so much for Greedy, RandomLS, and Cluster. Only Random algorithm goes up quickly as the number of users increases. Greedy method works on mallow's data sets showing that the running time increases relatively fast than other two methods, and has a nice GF result. We can see that the clustering method perform well both on running time and GF result. Varying the number of items, both running time and GF get relatively small value, which coincident with the real world data sets(Fig. 2).

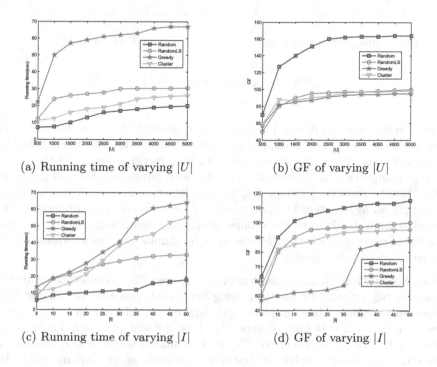

(a) Running time of varying $|U|$ (b) GF of varying $|U|$

(c) Running time of varying $|I|$ (d) GF of varying $|I|$

Fig. 2. Scalability on varying $|U|$ and $|I|$

The P-L data sets have a similar result with mallow's data sets which we cut it short. Figure 3 shows that all the proposed algorithms perform well with the scalability on varying $|l|$. From the result we obtained, it's obvious that the proposed Cluster and RandomLS have a relatively good performance regarding both running time and GF values. They both have fast convergence speed and small GF values which is what we seek for group formation on ranking data with related group recommendation methods.

6   Related Work

While no prior work has addressed the problem of group formation in the context of recommender systems on ranking data, we still discuss existing work that appears to be contextually most related.

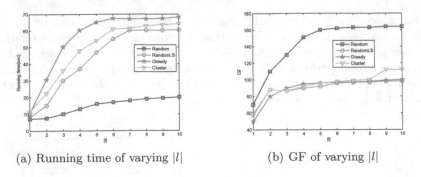

(a) Running time of varying $|l|$ (b) GF of varying $|l|$

Fig. 3. Scalability on varying $|l|$

Group Recommendation: There are two dominant strategies for group recommendations [6,8]. The first approach creates a pseudo-user representing the group and then makes recommendations to that pseudo-user, while the second strategy computes a recommendation list for each group member and then combines them to produce a groups list. For the latter, a widely adopted approach is to apply an aggregation function to obtain a consensus group preference for a candidate item. In these works, groups are created beforehand, either by a random set of users with different interests, or by a number of users who explicitly choose to be part of a group.

Team Formation: Team formation problems are often modeled using Integer Programming, or heuristic solutions using Simulated Annealing or Genetic Algorithms are designed. These problems are assignment problems. In general, group formation is not a matching or generalized assignment problem. There are no resources to match the users to. We need to match users to one another. In that sense, its closer in spirit to clustering However, as we demonstrate in the paper, a clustering algorithm which is agnostic to the group recommendation semantics is likely to perform poorly for purposes of minimizing the maximum disagreement between users.

Activity/Event Recommendation: These problems [11–13] recommend suitable activities and events to the communities (a set of users) with common interests. Again, our groups have a more explicit connotation (in the sense of having clearly defined satisfaction scores) than communities. One can potentially generate a graph of users based on a suitable notion of distance or similarity between users in terms of tastes and find communities.

7   Conclusions

To address the problem of group formation from group recommendation based on ranking data, four methods are proposed. Through deep insight experiment on the real world data sets and synthetic data sets, our proposed algorithms perform well. To our best knowledge, it is the first comprehensive discussion on this group formation, especially on ranking data.

References

1. Tong, Y., She, J., Ding, B., Wang, L., Chen, L.: Online mobile micro-task allocation in spatial crowdsourcing. In: ICDE, pp. 49–60 (2016)
2. Song, T., Tong, Y., Wang, L., She, J., Yao, B., Chen, L., Xu, K.: Trichromatic online matching in real-time spatial crowdsourcing. In: ICDE, pp. 1009–1020 (2017)
3. Crossen, A., Budzik, J., Hammond, K.J.: Flytrap: intelligent group music recommendation. In: IUI, pp. 184–185. ACM (2002)
4. Jameson, A., Smyth, B.: Recommendation to groups. In: Brusilovsky, P., Kobsa, A., Nejdl, W. (eds.) The Adaptive Web. LNCS, vol. 4321, pp. 596–627. Springer, Heidelberg (2007). https://doi.org/10.1007/978-3-540-72079-9_20
5. O'Connor, M., Cosley, D., Konstan, J.A., Riedl, J.: PolyLens: a recommender system for groups of users. In: Prinz, W., Jarke, M., Rogers, Y., Schmidt, K., Wulf, V. (eds.) ECSCW 2001. Springer, Heidelberg (2001). https://doi.org/10.1007/0-306-48019-0_11
6. Amer-Yahia, S., Roy, S.B., Chawlat, A., Das, G., Yu, C.: Group recommendation: semantics and efficiency. PVLDB 2(1), 754–765 (2009)
7. Senot, C., Kostadinov, D., Bouzid, M., Picault, J., Aghasaryan, A., Bernier, C.: Analysis of strategies for building group profiles. In: De Bra, P., Kobsa, A., Chin, D. (eds.) UMAP 2010. LNCS, vol. 6075, pp. 40–51. Springer, Heidelberg (2010). https://doi.org/10.1007/978-3-642-13470-8_6
8. Berkovsky, S., Freyne, J.: Group-based recipe recommendations: analysis of data aggregation strategies. In: RecSys., pp. 111–118 (2010)
9. Pizzutilo, S., De Carolis, B., Cozzolongo, G., Ambruoso, F.: Group modeling in a public space: methods, techniques, experiences. In: AIC., pp. 175–180 (2005)
10. Ntoutsi, E., Stefanidis, K., Nørvåg, K., Kriegel, H.-P.: Fast group recommendations by applying user clustering. In: Atzeni, P., Cheung, D., Ram, S. (eds.) ER 2012. LNCS, vol. 7532, pp. 126–140. Springer, Heidelberg (2012). https://doi.org/10.1007/978-3-642-34002-4_10
11. She, J., Tong, Y., Chen, L.: Utility-aware social event-participant planning. In: SIGMOD, pp. 1629–1643 (2015)
12. Tong, Y., She, J., Meng, R.: Bottleneck-aware arrangement over event-based social networks: the max-min approach. World Wide Web 19(6), 1151–1177 (2016)
13. She, J., Tong, Y., Chen, L., Cao, C.C.: Conflict-aware event-participant arrangement and its variant for online setting. IEEE Trans. Knowl. Data Eng. 28(9), 2281–2295 (2016)
14. Chen, J., Chen, X., Song, X.: Bidder's strategy under group-buying auction on the internet. IEEE SMC 32(6), 680–690 (2002)
15. Mallows, C.L.: Non-null ranking models. I. Biometrika 44(2), 114–130 (1957)
16. Plackett, R.L.: The analysis of permutations. Roy. Stat. Soc. 24(2), 193–202 (1975)

Taxi Call Prediction for Online Taxicab Platforms

Liefeng Rong[1](✉), Hao Cheng[1], and Jie Wang[2]

[1] SKLSDE Lab, Beihang University, Beijing, China
{rongliefeng,haocheng}@buaa.edu.cn
[2] Didi Research, Beijing, China
wangjiejacob@didichuxing.com

Abstract. The online taxi-calling services have gained great popularity in the era of sharing economy. Comparing with the traditional taxi service, the online taxi-calling service is much more convenient and flexible for passengers, because the taxi platform can provide detailed travel arrangements, transparent estimated price and flexible means of calling in advance. To understand the call willingness of passengers and to increase the quantity of orders, it is important to predict whether a request will be converted to a call order. In this paper, we study the problem of taxi call prediction, which is one of the important components of taxi-calling service system. To solve the problem, we propose a prediction framework, which uses some classification models with combining the basic features such as the order and meteorology information and the personlized historical profiles. Extensive experimental evaluations on real taxi record data from an online taxi-calling service platform demonstrated the effectiveness of our approach.

Keywords: Taxi call prediction · Personalized · Sharing economy

1 Introduction

Sharing economy is getting more and more popular. Online taxi-calling apps/platforms such as Didi, Uber and Lyft have emerged as a novel and popular means to provide transportation service via mobile apps. Comparing with the traditional taxi service, subway and bus, the online taxi-calling service is much more convenient and flexible for the passengers. A large number of taxi demand requests are generated routinely every day. For example, more than one million taxi-calling requests are generated in Beijing per day on Didi, the largest online taxi-calling service provider in China. One question arises: how many call orders will be generated among those requests? More concretely, we want to predict whether a request will be converted to a call order. Imagine the following scene. To call a taxi, a passenger Andy simply types in her desired source and destination in the app and sends the request to the platform, who instantly feeds back a price and other detailed travel information (e.g. the estimated distance and

© Springer International Publishing AG 2017
S. Song et al. (Eds.): APWeb-WAIM 2017 Workshops, LNCS 10612, pp. 214–224, 2017.
https://doi.org/10.1007/978-3-319-69781-9_21

the discount). Then Andy determines to click the call option button in the app to hailing a taxi, or she leaves the calling page because maybe she can not afford the fare. If Andy calls, the platform assigns a close-by driver to take the order. To understand the call willingness of passengers and to increase the quantity of orders in similar scenes, it is important to predict whether a taxi request will be converted to an order. We say this problem as Taxi Calling Prediction (TCP) problem. We denote the ratio between the number of requests and actual orders during a period (like a day) in a district as Taxi Calling Ratio (TCR), which represents the average probability of requests converting to orders.

Taxi calling prediction information benefit the platforms in triple ways. (i) *Evaluating price mechanisms.* TCR reflects the willingness of passengers to travel by taxi after the platform adopts new price mechanisms and discount strategies. (ii) *Expanding potential market.* By observing historical data, the platforms can discover times and regions with low TCR, which have room for growth. (iii) *Customizing individual needs.* The platforms can learn passenger behaviors from user historical records, and then choose the most suitable prices and the best travel scheme for users. Therefore, TCP is a foundational issue in large-scale online taxi industries.

To solve the problem of TCP, we propose a unified TCP framework with some classification models and spatio-temporal features. Figure 1 illustrates the overview of the framework. We first investigate multiple real-world datasets including taxi records and meteorology. Then we extract four types of basic features over time, space, money and meteorology domains to train models as a basic version. And we present an advanced version by generating user personalized historical features based on the understandings of the business logics of online taxicab platforms.

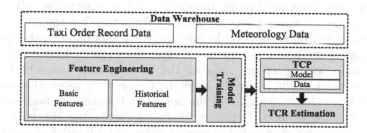

Fig. 1. An overview of the framework.

Contributions. To the best of our knowledge, this is the first effort on the problem of taxi call prediction for large-scale online taxicab platforms. We transform the overhead of sophisticated model redesign into feature engineering, and apply three state-of-the-arts classification models for training and testing. We conduct extensive experiments on two real datasets from the largest online taxicab platform in China. The experimental results show that our prediction models perform well.

The rest of the paper is organized as follows. We describe the details of the real-world large datasets used in our study in Sect. 2, and introduce our feature engineering in Sect. 3. The training models are proposed in Sect. 4. The experimental setup and results are presented in Sect. 5. We discuss some related work in Sect. 6 and conclude the paper with a summary in Sect. 7.

2 Data Description

This section introduces the datasets used for TCP, including original taxi order records and meteorological information. Our datasets were collected in two metropolises in China (Beijing and Nanjing). Due to the space limit, we only describe the Beijing dataset.

2.1 Taxi Order Record Data

The original taxi order records of the Beijing dataset are sampled in proportion from Didi. The Beijing raw dataset contains 23,851,235 original taxi order records from July 1, 2016 to December 31, 2016. Each record consists of a user ID, a time stamp, locations of the origin and destination, an estimated distance, an estimated price and the discount information. The user ID, time stamp, origin and destination are submitted by users, and the estimated distance, the estimated price and the discount information are calculated by the online taxicab platform based on the information submitted by users. And each record has a label which denotes weather the request turns into an actual order.

Figures 2a–d depict the distributions of the taxi order records. Figure 2a plots the distribution of the normalized number of taxi order records per day. The blue polyline denotes the request order, the green polyline denotes the call order and the red polyline represents the trend of TCR. We notice that the number of taxi request orders exhibit weak periodic patterns across six months, the number of taxi call orders are positively correlated with the request, however, TCR also shows weak periodic patterns. Because the taxi demand and actual call will can be influenced by the relatively aperiodic factors such as weather and price. Figure 2b shows the distribution of the normalized number of taxi orders with respect to different estimated distances. Same as Fig. 2a, the blue curve and the green curve represent the number of request order and call order respectively, the red curve denotes the TCR. As shown, after 3 Km the TCR decreases with the distance increases, indicating that long rides have low probability to be called. Figure 2c demonstrates the distribution (blue curve) and the TCR (red curve) of discounts, from which we see that most taxi orders get discounts within [0.6,1.0] and discounts within [0.25,0.6] have high TCR.

2.2 Meteorology Data

The meteorology data in Beijing is collected from an meteorology website of Chinese government and the range of the data is from July 1 to December 31 in 2016.

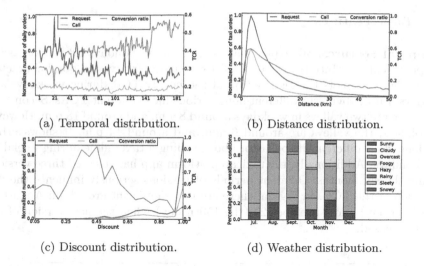

(a) Temporal distribution.   (b) Distance distribution.

(c) Discount distribution.   (d) Weather distribution.

Fig. 2. (a) Normalized numbers of daily taxi orders in six months; (b) distribution of estimated taxi-trip distances; (c) distribution of discounts; (d) monthly distribution of weather information. (Color figure online)

Each record consists of a timestamp, weather condition, temperature, humidity, air quality and wind speed and the records are updated once an hour. There are 8 different weather conditions in the data and the distribution of weather in each month is shown in Fig. 2d. We can observe that the proportion of haze ascends from July to December while the frequency of rainy days falls. The air quality is classified in 6 levels: excellent, good, lightly polluted, moderately polluted, heavily polluted and severely polluted. 6% of the meteorology records are imcomplete or empty. For the missing temperature, wind speed, and humidity, we adopt the average of the values of the previous hour and the next hour, while we use the same value in the previous hour to represent the missing weather condition and air quality.

3   Feature Engineering

In this section, we select the proper feature sets for TCP via feature engineering. Feature engineering is the process of using domain knowledge of the data to create features for representing the humans understanding about influence factors of complicated problems. Specifically, for our TCP problem, we first consider taxi order record features including temporal, spatial, monetary and meteorological as a basic version. And then we propose an advanced version combining the basic taxi order record features and personalized historical features.

3.1   Basic Features

Temporal Features. We exploit day of week, hour, holiday and version as the temporal features. Intuitively, the TCR exhibits distinctive temporal characteristics. Figure 3a plots the distribution of the hourly TCR during weekdays and holidays. First, obviously, different hour periods have different TCRs. Second, we find there are two valleys in weekdays around 8 a.m. and around 18 p.m. However in holidays, the valleys are around 6 a.m. and around 14 p.m. It shows urban residents have different life behavior about calling a taxi online during weekdays and holidays. Additionally, the taxicab platform app has released three versions during the six months. Different periods of versions seriously influence the distribution of the TCR. In Fig. 2a, there are tow significant growth on August 27 (the 58th day) and November 27 (the 150th day) when the new version of the app is released.

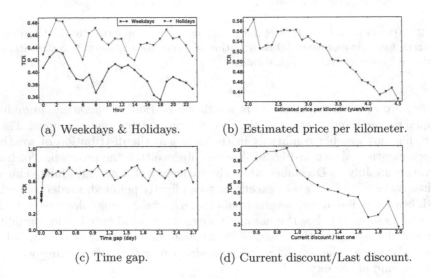

(a) Weekdays & Holidays.   (b) Estimated price per kilometer.

(c) Time gap.   (d) Current discount/Last discount.

Fig. 3. Distribution of features.

Spatial Features. The Orange curve in Fig. 2b demonstrates the distribution of the TCR with respect to different estimated distances, from which we see that the TCR decreases as the estimated distance increased. We adopt estimated distance and the square of estimated distance as the spatial features.

Monetary Features. We use monetary features of estimated price, estimated price per kilometer and discount because they are very important factors which affect the incentives of taxi-calling users, and consequently, the TCR. Figure 3b shows the relationship between the estimated price per kilometer and the TCR for orders under 4 km. We notice that the estimated price per kilometer exhibits negative correlation with the TCR. Figure 2c plots the relationship between the average discount and the TCR. It can be observed that when the platform provides more discounts, the TCR tends to increase in general.

Meteorological Features. Meteorological information such as weather can be an important consideration to alter transportation modes, and accordingly, an impacting factor of TCR. For example, in Fig. 2a, the valley in the red circle is July 20, when there was a torrential rain, leading to a surge of demands and a decline of TCR. Other than weather condition, we also use temperature, humidity and air quality as the meteorology features for the similar reasons.

3.2   Historical Features

Intuitively, we collect three referential features of historical request number, historical call number and historical TCR as a apart of historical features for TCP. Additionally, we choose request time gap, last request status and last request discount as the other historical features.

As shown in Fig. 3c, short time gap between current request order and last one relates with a low TCR. Conversely, time gap more than an hour shows has a high and windless TCR for the reason that some price sensitive users may send request orders during a short period of time to find a lower price. And an office worker who is not a price sensitive user may send a taxi call order every weekday at 18p.m. to go home, and her/his request time gap is long. For one passenger, each request has three status: the last request is a call order, the last request is not a call order and has no last request. Table 1 demonstrates the influence of the different status of users' last request to the current TCR, from which we can see that if the last request's status is a call order, the current TCR is higher. We also plots the distributions of the ratio between the discount of current request and the last one in Fig. 3d. We count the ratio from 0.5 to 2.0, and the results shows that when the platform provides less discount for the current request order, the TCR tends to decrease, indicating passengers are discount-sensitive in general.

Table 1. TCR of the different status of last requests.

Status	Request count	Call count	TCR
None	2696	1417	52.6%
Not a call order	48599	20158	41.5%
A call order	37545	20777	55.3%

4   Prediction Model

As proposed in Sect. 1, Fig. 1 illustrates the overview of the framework. After data description and feature engineering, we present prediction models in this section. We adopt three popular classification models including Logistic

Regression, Gradient Boosting Decision Tree and Random Forest. We don't focus on the model methods, hence we briefly introduce them. And we evaluate our train models with four metrics.

Training methods

Logistic Regression (LR): A frequently-used model using for regression or classification.

Gradient Boosting Decision Tree (GBDT): Gradient Boosting Decision Tree is a powerful ensemble method which is widely used in data mining applications. In our experiment, we use a fine-tuned and efficient GBDT implementation XGBoost.

Random Forest (RF): Random Forest is another widely used ensemble method which offers comparable performance with GBDT. We use the RF implementation from the scikit-learn library.

Evaluation metrics. We use AUC, precision, recall and F-score computed with test set to evaluate the performances of different methods. AUC is the area under the ROC curve. The precision is the number of correctly identified positives divided by the number of identified positives instances. The recall is the number of correctly identified positives divided by the number of all positive instances in the test set. Then, the F-score is defined as:

$$Fscore = 2 \times \frac{precision \times recall}{precision + recall} \tag{1}$$

5   Experiments

In this section, we present experimental results with our proposed framework. We first describe the details of our dataset and the experimental setting. We summarize the experiments results to compare the features of two versions and the three classification models. We also evaluate the usefulness of feature extracted in Table 2.

5.1   The Experimental Setup

We evaluate the performance of our scheme on both the Beijing and the Nanjing datasets. We chronologically order each dataset and use the first 3/4 for training and the remaining 1/4 for testing.

5.2   Overall Results

Table 2 summarizes the results of the three compared models with respect to the four evaluation metrics in different feature types and different cities. From the results, we observe that comparing with the basic version, the advanced version has a huge improvement in every evaluation metrics. For example, in

Table 2. The performance comparison.

City	Feature type	Method	Precision	Recall	F-score	AUC
Beijing	Basic	LR	0.531563	0.449135	0.486885	0.673521
		RF	0.619659	0.467034	0.532628	0.726518
		GBDT	0.616175	0.462558	0.528429	0.722539
	Basic + Historical	LR	0.739809	0.697256	0.717902	0.862802
		RF	0.801062	0.711289	0.753511	0.900445
		GBDT	0.797747	0.681731	0.735190	0.884008
Nanjing	Basic	LR	0.501520	0.386137	0.436329	0.596301
		RF	0.571032	0.426101	0.488033	0.651039
		GBDT	0.566614	0.417167	0.480539	0.644363
	Basic + Historical	LR	0.738614	0.707438	0.722690	0.852067
		RF	0.797806	0.744752	0.770367	0.900320
		GBDT	0.809891	0.809891	0.769572	0.900224

Beijing data, the AUC of advanced version with random forest model is 23.9% higher than the basic version. It shows that developing the personalized historical features can improve the accuracy of TCP. We can also make the following observations by comparing the three model. The RF performs poorly on both datasets and GBDT and RF both have outstanding performance. RF outperforms GBDT in almost all the metrics on the two datasets. The only exception is the precision metric on the Nanjing dataset in advanced version, where GBDT yields slightly higher precision. In summary, using random forest model with the basic features adding personalized historical features generally outperforms best in TCP.

Table 3. The top 10 features ranked by IGR.

Rank	Feature	Importance
1	Historical call number	0.221967
2	Discount	0.103552
3	Estimated price	0.082832
4	Request time gap	0.077289
5	Estimated time of arrival	0.040514
6	Estimated distance	0.040212
7	Historical request num	0.033359
8	Historical request distance	0.029758
9	Historical request status	0.017493
10	Version	0.006368

5.3   Feature Contribution Analysis

We further examine the contributions of each individual feature. Specifically, we leverage the widely-used information gain ratio (IGR) as a metric to determine which of the features are the most important. Table 3 shows the top 10 features ranked by their IGR (i.e., higher IGR means greater importance). It can be observed that the most contributive feature is the personalized historical TCR and there are 6 personalized historical features in top 10, demonstrating the necessity to develop personalized historical features.

6   Related Work

The TCP problem we study in this paper is closely related to the following two categories of research.

6.1   Taxi Demand Prediction

The taxi demand prediction studies the problem of forecasting the demands in every pick up location, which can further guide and optimize the taxi dispatching [11] and task assignment in location-based services [9,10]. Moreira-Matias et al. [7] design a model to predict the number of future services at a given taxi stand, where the GPS traces and event signals are transformed into a time series of interest as both a learning base and a streaming test framework. Recently, Tong et al. [8] propose a simple unified linear regression approach with massive combinational features to estimate real-time taxi demands, which successfully uses feature engineering methods to improve the accuracy of the taxi demand prediction. Above works only studied the demand prediction, and there are some research predicting the the supply-demand. Anwar et al. [1] combine the trajectories of taxi and flight arrival data to predict the unmet taxi demands, which means the gap between taxi demands and potential supply of taxicabs at airports. Dong et al. [12] predict the equilibrium of the supply-demand, and use environment data such as the weather or traffic conditions to enhance the prediction accuracy. Both demand prediction and supply-demand prediction are regression problems essentially, they predict the number of taxi demand or demand gap in a period of time at an area, hence it is impossible to extend works on demand prediction to TCP.

6.2   Ad Click Prediction

Predicting ad clickthrough rates (CTR) is a massive-scale learning problem that is central to the online advertising industry [6]. And traditional ad click prediction usually appears in two forms: sponsored search and social network. Thore et al. [3] describe a Bayesian click-through rate prediction algorithm, which is based on a probit regression model, used for Sponsored Search in Microsoft's Bing search engine. Haibin et al. [2] present a framework for the personalization of click models by developing user-specific and demographic-based features in sponsored search. In social network, Xinran et al. [4] introduce a model which combines decision trees with logistic regression for the ad click prediction problem in facebook advertising systems. Cheng et al. [5] propose a learning-to-rank method for the problem of click-through prediction for advertising in Twitter timeline. We stress that the CTR prediciton problem isn't involving money trading. Moreover, none of these work studied the environment data such as the weather or spatial information.

7   Conclusion

In this paper, we introduce a realistic problem called taxi call prediction (TCP) problem for online taxicab platforms, which attempts to predict whether a taxi request will be converted to a call order. We propose a prediction framework, which using three classification models including LR, GBDT and RF, and combining the basic features and the personlized profiles. Extensive evaluations on two large-scale datasets from an industrial online taxicab platform validate the effectiveness of our approach. We envision our experiences of taxi call prediction can serve as an insightful reference for other purchase prediction problems of shared economy application services with spatio-temporal features.

References

1. Anwar, A., Volkov, M., Rus, D.: Changinow: a mobile application for efficient taxi allocation at airports. In: ITSC, pp. 694–701 (2013)
2. Cheng, H., Cantú-Paz, E.: Personalized click prediction in sponsored search. In: WSDM, pp. 351–360 (2010)
3. Graepel, T., Candela, J.Q., Borchert, T., Herbrich, R.: Web-scale Bayesian click-through rate prediction for sponsored search advertising in microsoft's bing search engine. In: ICML, pp. 13–20 (2010)
4. He, X., Pan, J., Jin, O., Xu, T., Liu, B., Xu, T., Shi, Y., Atallah, A., Herbrich, R., Bowers, S., Candela, J.Q.: Practical lessons from predicting clicks on ads at Facebook. In: ADKDD, pp. 5:1–5:9 (2014)
5. Li, C., Lu, Y., Mei, Q., Wang, D., Pandey, S.: Click-through prediction for advertising in twitter timeline. In: SIGKDD, pp. 1959–1968 (2015)
6. McMahan, H.B., Holt, G., Sculley, D., Young, M., Ebner, D., Grady, J., Nie, L., Phillips, T., Davydov, E., Golovin, D., et al.: Ad click prediction: a view from the trenches. In: SIGKDD, pp. 1222–1230 (2013)
7. Moreira-Matias, L., Gama, J., Ferreira, M., Mendes-Moreira, J., Damas, L.: Predicting taxi-passenger demand using streaming data. IEEE Trans. Intell. Transp. Syst. 14(3), 1393–1402 (2013)
8. Tong, Y., Chen, Y., Zhou, Z., Chen, L., Wang, J., Yang, Q., Ye, J., Lv, W.: The simpler the better: a unified approach to predicting original taxi demands based on large-scale online platforms. In: SIGKDD, pp. 1653–1662 (2017)
9. Tong, Y., She, J., Ding, B., Chen, L., Wo, T., Xu, K.: Online minimum matching in real-time spatial data: experiments and analysis. Proc. VLDB Endow. 9(12), 1053–1064 (2016)

10. Tong, Y., She, J., Ding, B., Wang, L., Chen, L.: Online mobile micro-task allocation in spatial crowdsourcing. In: ICDE, pp. 49–60 (2016)
11. Tong, Y., Wang, L., Zhou, Z., Ding, B., Chen, L., Ye, J., Xu, K.: Flexible online task assignment in real-time spatial data. Proc. VLDB Endow. **10**(11), 1334–1345 (2017)
12. Wang, D., Cao, W., Li, J., Ye, J.: DeepSD: supply-demand prediction for online car-hailing services using deep neural networks. In: ICDE, pp. 244–254 (2017)

SDMA 2017

Data Gathering Framework Based on Fog Computing Paradigm in VANETs

Yongxuan Lai[1](✉), Lu Zhang[1], Tian Wang[2], Fan Yang[3], and Yifan Xu[1]

[1] Department of Software Engineering, Software College,
Xiamen University, Xiamen 361005, China
laiyx@xmu.edu.cn, 1042842053@qq.com, 525608308@qq.com
[2] College of Computer Science and Technology,
Huaqiao University, Xiamen 360000, China
wangtian@hqu.edu.cn
[3] College of Aerospace Engineering, Xiamen University, Xiamen 361005, China
yang@xmu.edu.cn

Abstract. Vehicular nodes are equipped with more and more sensing units, and large amount of sensing data are generated. This makes the data gathering and monitoring is a challenging problem in VANETs. In this paper we first present an sensing and data gathering framework through the concept of fog computing in VANETs, then we propose an event-based data gathering scheme based on this framework that adaptively adjusts the threshold to upload suitable amount of data for decision making, while at the same time suppress unnecessary message transmissions. Preliminary experiments demonstrate the effectiveness of the proposed algorithm in vehicular sensing applications.

Keywords: Fog nodes · Data gathering framework · VANETs

1 Introduction

One key and challenging issue in VANETs [1,4] is the vehicular sensing and data gathering. Vehicular nodes are equipped with more and more sensing units, and large amount of sensing data such as GPS locations, speeds, video clips, or even chemical emissions are generated [12]. These data are shared or uploaded as an input for applications aiming at more intelligent transportation, emergency response, and reducing pollution and fuel consumption. In other words, cooperative urban sensing is at the heart of the intelligent and green city traffic management. The key components of the platform will be a combination of pervasive vehicular sensing system and a central control and analyzing system, where data gathering is a fundamental component that bridges the two systems.

However, the data gathering and monitoring is also a challenging issue in VANETs [9,11]. Firstly, VANETs differ from other Mobile Ad hoc Networks

Supported by the Natural Science Foundation of China (61672441, 61303004, 61572206), the National Key Technology Support Program (2015BAH16F01), the State Scholarship Fund of China Scholarship Council (201706315020).

S. Song et al. (Eds.): APWeb-WAIM 2017 Workshops, LNCS 10612, pp. 227–236, 2017.
https://doi.org/10.1007/978-3-319-69781-9_22

by its own characteristics. The network scale could be large, and the vehicular nodes are with high computational ability and no power constraints. Also, nodes are limited to road topology while moving, and under various road conditions and high moving speed the network usually suffers rapid topology and density changes. The communications are usually fragmented and intermittent-connected. Secondly, the vehicular sensed data is in large amount and characterized as continuous generation. The sensed data should be filtered and pre-processed before being shared or uploaded; data filtering technologies that tailored to the VANET environment are highly needed.

One potential solution to efficient sensing and data gathering in VANETs is through fog nodes, which is the concept of fog computing [2,4,17] that extends the traditional cloud computing paradigm to the edge in VANETs. The fog nodes, e.g. RSUs, are able to provide computation, storage, and networking services between the vehicular nodes and ITS central servers, which brings about the chances and opportunities for the optimization of data or event gathering process in VANETs. In this paper, we propose an data gathering framework called DAFOC (Data gAthering based on FOg Computing) in VANETs. The framework is consisted of three layers, denoted as the local layer, the fog layer, and the cloud layer. Then based on DAFOC, we proposed an event-based data gathering scheme that adaptively adjusts the threshold to upload suitable amount of data for decision making, while at the same time suppress unnecessary message transmissions. Preliminary experiments demonstrate that effectiveness of the proposed algorithm in vehicular sensing applications.

The rest of the paper is structured as follows: Sect. 2 describes the related work; Sect. 3 presents the detailed description of the DAFOC framework; Sect. 4 presents an event-based data gathering algorithm based on DAFOC; finally, Sect. 5 concludes the paper.

2   Related Work

Vehicles could be viewed as powerful mobile sensors, at this section we briefly survey some related works on data/event gathering and fog related research on VANET.

Lee et al. [13] proposed the MobEyes system for proactive urban monitoring. The system exploits the vehicle mobility to opportunistically diffuse concise summaries of the sensed data, and it harvests these summaries and builds a low-cost distributed index of the stored data to support various applications. Palazzi et al. [15] proposed a delay-bounded vehicular data gathering approach, which exploits the time interval to harvest data from the region of interest satisfying specified time constraints, and properly alternates the data muling and multi-hop forwarding strategies. Yet the solution has to be integrated with a time-stable geocast protocol for the query propagation.

Besides minimizing the delay from source to destination, event monitoring and data gathering protocols also try to minimize the consumed resources while ensuring that the collected information meets certain maximum delay requirements. Delot et al. [3] proposed a pull-based data gathering called GeoVanet.

It adopts a DHT-based model to identify a fixed geographical location where a mailbox is dedicated to the query, so users are able to send queries to a set of cars and find the desired information in a bounded time. Paczek [16] introduced a method of selective data collection for traffic control applications. The underlying idea is to detect the necessity of data transfers on the basis of uncertainty determination of the traffic control decisions, and sensor data are transmitted from vehicles to the control node only at selected time moments. Lindgren et al.

For the research on vehicular and clouds, Eltoweissy et al. [5] for the first time coined the term of Autonomous Vehicular Clouds (AVC), where a group of largely autonomous vehicles whose corporate computing, sensing, communication, and physical resources can be coordinated and dynamically allocated to authorized users. Hussain et al. The concept of VANET Cloud, however, is highly related to the fog computing, which extends traditional cloud computing paradigm to the edge. Bonomi et al. [2] defined the characteristics of fog computing and its role in the IoT. They emphasized the fact that the fog brings new elements to the realm of Internet of Things through reduction of service latency and improvement of QoS. Vaquero et al. More recently, Kai et al. [6] gave a survey on some opportunities and challenges related to the context of fog computing in VANETs.

3 DAFOC Framework

We assume each vehicle, v_i, monitors the road condition and surrounding environment through periodical sensing (Fig. 1). It generates pieces of data and sends them to the roadside units (RSUs) through one hop vehicle to infrastructure (V2I) or multi-hop vehicle to vehicle (V2V) transmissions. RSUs also serve as the *fog nodes* to provide computation, storage, and networking services between

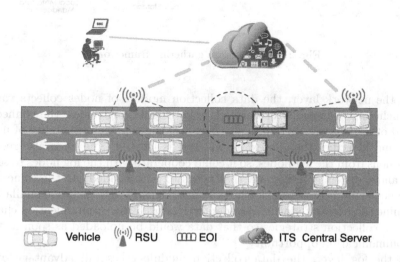

Fig. 1. Illustration of a vehicular sensing system.

the vehicular nodes and ITS (Intelligent Transmission System) clouds. An RSU would initiate an event checking procedure to evaluate the received data, which might lead to the upload of more detailed data to the base station when it considers necessary. The detailed data are uploaded within constrained time delay to the central ITS server. The server, due to its powerful computational resource and global knowledge of the road and network, would then make the final decisions based on further data analysis.

Figure 2 presents the three-tier data collection architecture called DAFOC, which includes the network layer, the fog layer, and the cloud layer. The framework would make full use of the computing, storage and communication capabilities of the fog nodes, and their positional relevance and real-time at the edge of the network. It is an underlying data gathering framework for various VANET applications that consume the sensing data.

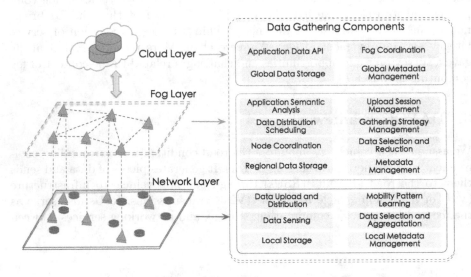

Fig. 2. DAFOC data gathering framework.

At the network layer, the data collection module of nodes collects various data including the encounter data with the neighboring nodes, the connection log and other network metadata. It calculates the contact probability of nodes, learns and identifies the context in which it is located. The node also receives instructions from the fog nodes, so it could cooperate with the fog nodes to select, filter, and reduce the local data that needs to be uploaded. At the same time, the vehicular nodes would calculate the data density and the node similarities, determine the network status and adaptively switch the communication channel and data collection strategies so that data would be uploaded as soon as there is a communication opportunity.

At the fog layer, the data collection module takes full advantage of the processing capacity and location characteristics of the fog nodes to optimize the efficiency and effectiveness of network data collection. The fog nodes communicate with the cloud, parse the semantic information of the application and obtain the data collection requirements. For each fog node, it receives the sensing data and metadata from the network layer and uploads the collected data to the cloud. The fog node will calculate the data selection and reduction strategy between nodes according to the application semantics and network status. The network layer is also optimized according to the network condition evaluation model, and the data upload and dispatch will be coordinated with each node to avoid transmission conflicts and message collisions. By adopting an adaptive data collection strategy, and through the construction of "upload session", nodes could switch quickly and seamlessly when uploading the data at different environments. At the same time, the fog node will work with the cloud, adopting an "fog - cloud" collaborative storage strategy to provide a unified, efficient, real-time data services for VANET applications.

At the cloud level, the data collection module will provide interfaces for the VANET application to store the gather data from the fog layer. The cloud manages the global metadata information of the network, and provides APIs for each application and receives processing requests from the fog nodes. At the same time, the cloud would collaborate with the fog nodes to generate a comprehensive data collection framework.

4   Data Gathering Based on DAFOC

At this section we propose an event-driven data gathering scheme based on DAFOC. We assume the sensing operators equipped in vehicular nodes are roughly classified into two types/modes: the low cost sensing (LCS) mode and high cost sensing (HCS) mode. The former requires much less sensing and calculation power than the latter. Each piece of sensed data is denoted as a tuple: $d(id, v, \alpha, ts, mode)$, where id is a unique identification of the sensing node, v is the sensed reading, α indicates the confidence of the true reading, ts is the timestamp, and $mode$ is the mode of sensing, which is either LCS or HCS.

Figure 3 depicts the overall phrases of the data gathering scheme. There are 5 steps as follows:

1. **Data Monitoring:** Nodes work on the LCS mode, they sense the environment and generate the data. The data are of relatively low generation rate

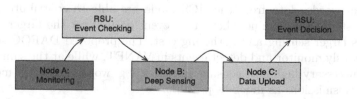

Fig. 3. Event-driven data gathering based on DAFOC.

and confidence, which are then uploaded to the RSU through V2I or V2V communications.

2. **Event Checking:** RSU checks the confidence/probability of an event, and it initiates an event checking procedure when the event probability is high.
3. **Deep Sensing:** Some nodes are elected to transit to "deep sensing", where they work on the HCS mode and sense more accurate data about the environment.
4. **Data Upload:** The data generated at HCS mode are uploaded to the RSU for final event verification. Data could be uploaded directly, forwarded to neighboring nodes, or wait until encountering a new RSU. Nodes would adaptively decide their strategy for data upload based on whether they are within the coverage of an RSU or encountered nodes.
5. **Event Decision:** The ITS cloud processes the gathered data and makes decision about the events. Some data is archived at the cloud, and some feedbacks are sent back to the RSUs.

The monitoring procedure at step 1 runs as background process with low cost, while still be able to alert possible events. Vehicular nodes would upload the data of event at the deep sensing step, while suppress the message communications at the monitoring step. The event of interest (EOI) under monitoring is denoted as e, and the vehicular nodes near the EOI would sense the event and generate some data. There is a mapping of correlation between the set of gathered data D and event e. The correlation is indicated by a function denoted as $\hbar(d, e)$.

$$\hbar(d, e) = \{p(e), \alpha\}, \quad p(e) \in \{1, -1\}, \alpha \in [0, 1] \tag{1}$$

$p(e)$ is the prediction of event, where 1 denotes the event occurs and -1 denotes otherwise; α is the confidence of the prediction. And the *weight* of d on event e could be further defined as:

$$w(d, e) = p(e) \times \alpha \tag{2}$$

The weight is highly related to the sensing operator and the monitoring applications. If the weight of data, e.g. $w(d, e)$ is close to 1, e is likely to occur; if it is close to -1, e is unlikely to occur; if it is close to 0, the occurrence of the event is unclear due to lack of confidence. Given the type of sensing operator and data, the weight of data is used as the main input for the event monitoring and detection. Also, we assume the data gathered at HCS mode has much larger weight than that at LCS mode:

$$w(d_1, e) \ll w(d_2, e), \quad d_1.mode = LCS, d_2.mode = HCS \tag{3}$$

In other words, data from the HCS mode are able to contribute more for the ITS system to determine whether an event occurs, yet the larger data size also incurs larger sensing and gathering cost. The proposed DAFOC scheme is to continuously monitor and detect events in VANET, while at the same time to reduce unnecessary high cost sensing activities, e.g. working in HCS mode, and message transmissions as much as possible.

In the following subsections, we present the detailed description of the main steps of the algorithm.

4.1   Data Monitoring

For an event monitoring and detection application, each node senses the data according to the monitoring command when it moves within the area of RSU. Each node works at the LCS mode and generates data. The weight of data d on event e is $w(d, e)$, and d is reported to the RSU if its weight is larger than a predefined threshold:

$$w(d, e) \geq \tau_0 \tag{4}$$

As mentioned at Sect. 3, we assume the weight of data is mainly decided by its corresponding cost, where data with lower cost have smaller weight on detecting an event. For example, photos with lower resolution are less effective for the detection a roadside accident than those of higher resolution. Yet they incur less sensing and transmission cost, and might trigger the gathering of detailed data for further analysis.

4.2   Event Checking

Vehicular nodes send their readings to the RSU, so RSU is able to calculate the weight of the received readings. For a set of readings D, the weight is defined as $w(D)$:

$$w(D) = \frac{\sum_{d \in D} w(d)}{T_D}, \quad T_D = max(d_1.ts - d_2.ts), \quad d_1, d_2 \in D \tag{5}$$

Given event e, the weight of data $w(d, e)$ is denoted as $w(d)$ for simplicity. Here T_D is defined as the time span of the readings in D. $w(D)$ is actually the weight of data scattered along the time interval. If the weight for a specific event within the time window is greater than threshold τ_1:

$$w(D) \geq \tau_1 \tag{6}$$

the RSU would broadcast an event checking procedure to all nodes within its covered area. Here we adopt a threshold-based event checking strategy. If a threshold is broken, it means an event is likely to occur on the road, and a further checking is required. Threshold τ_0 and τ_1 are two key parameters for our algorithm, they strike a balance between the number of uploaded readings and the cost of event checking procedures.

When an RSU initiates an event checking procedure, it periodically broadcasts an "event checking command" thin its covered area, where some passing-by nodes would transit to the HCS mode. However, working on the HCS mode is expensive. It needs more computing and energy resources to sense more accurate data, and also costs more transmissions to upload the data. Therefore it is crucial for DAFOC to selectively pick up a part of the passing-by nodes to work on this mode.

A node creates a timer when it receives an event checking command from an RSU for the first time. When the timer is fired, the node transits to the HCS mode. The timer is denoted $tm(\beta)$, where β is the delayed interval and defined as follows:

$$\beta = \begin{cases} BI * \frac{\tilde{t}}{\tilde{t}-t}, & t \leq \tilde{t} \\ BI * \delta, & t > \tilde{t} \end{cases} \quad BI < \tilde{t}, \ \delta \in (0,1) \tag{7}$$

where BI denotes the time interval of the broadcasting of *checking-event* messages at RSU, $\tilde{t}$ is the average duration of a node moving through the RSU coverage area, t is the elapsed time when a node enters the coverage area, and δ is a random number. Nodes enter the RSU covered area in sequential order, and the node that newly enters the area would have the smallest delay interval according to Eq. 7. It would have its timer tm fired and accept the command of working on HCS mode with less delay. Also, the broadcasting interval BI is set to be smaller than the average pass duration $\tilde{t}$, so when a node goes through the RSU coverage area it would receive at least one broadcasting message. At extreme cases when the elapsed time is larger than $\tilde{t}$, e.g. traffic jam, the delay β would be randomly set and nodes are randomly elected to work on the HCS mode.

If a node accepts the *checking-event* command, it would transit to HCS mode and send an *accept-hcs* message to the RSU, where other nodes overhearing this message would cancel their timers. So only a smaller part of the nodes would join the event checking procured and some nodes would still work on the LCS mode. For the RSU, it periodically broadcasts the event checking command if it does not receive an *accept-hcs* command from passing-by nodes. When it receives the *accept-hcs* message, it would broadcast an *on-checking* command, and other nodes that receive this message would cancel their timers for accepting the event checking command.

4.3   Data Upload

Nodes that accept the event checking command would transit to deep sensing for a period of time, and relatively larger size of data would be generated on the HCS mode. The data are uploaded to the RSUs, and then routed to the ITS system for event analysis as RSUs are inter-connected through wired networks or internet.

Here we borrow the idea of "store-carry-forward" in delay torrent network [10] for the data uploading and forwarding. When a node, e.g. s', receives data from another node, s' is responsible for the data uploading. s' might upload the data when it is within an RSU coverage, or forward it to another node. The calculation is mainly based on the expected time interval to reenter an RSU coverage area, as well as the expected contact duration between two encountered nodes, which are learned from the network meta-data and historical data. Each node would predict their time left to the next RSU ($et(s)$) based on their routes or historical trajectories, as well as contact histories with RSUs. The expected contact duration ($cd(s, s')$) between two encountered nodes could also be estimated based on their direction, speed, and trajectory records. These metadata and parameters are common input for the routing protocols for VANET. Readers could refer to [4] for further discussions.

5 Conclusions

In this paper we proposed an sensing and data gathering framework called DAFOC based on the concept of fog computing in VANETs. Then an event-based data gathering scheme is presented, which adopts a two-level threshold strategy to suppress unnecessary data upload and transmission. The proposed algorithm could effectively reduce the transmission cost when gathering the sensed data, while at the same time suppress unnecessary message transmissions. For the future work, we are going to extend the DAFOC framework and study its impact on various VANET applications with more detailed experiments.

References

1. Al-Sultan, S., Al-Doori, M.M., Al-Bayatti, A.H., Zedan, H.: A comprehensive survey on vehicular ad hoc network. J. Netw. Comput. Appl. 37, 380–392 (2014). http://www.sciencedirect.com/science/article/pii/S108480451300074X
2. Bonomi, F., Milito, R., Zhu, J., Addepalli, S.: Fog computing and its role in the internet of things. In: Proceedings of 1st edn. of the MCC Workshop on Mobile Cloud Computing, pp. 13–16. ACM (2012)
3. Delot, T., Mitton, N., Ilarri, S., Hien, T.: Decentralized pull-based information gathering in vehicular networks using GeoVanet. In: 2011 IEEE 12th International Conference on Mobile Data Management, vol. 1, pp. 174–183. IEEE (2011)
4. Dua, A., Kumar, N., Bawa, S.: A systematic review on routing protocols for vehicular ad hoc networks. Veh. Commun. 1, 33–52 (2014). Elsevier
5. Eltoweissy, M., Olariu, S., Younis, M.: Towards autonomous vehicular clouds. In: Zheng, J., Simplot-Ryl, D., Leung, V.C.M. (eds.) ADHOCNETS 2010. LNICSSITE, vol. 49, pp. 1–16. Springer, Heidelberg (2010). doi:10.1007/978-3-642-17994-5_1
6. Kai, K., Cong, W., Tao, L.: Fog computing for vehicular ad-hoc networks: paradigms, scenarios, and issues. J. China Univ. Posts Telecommun. 23(2), 56–96 (2016)
7. Keränen, A., Ott, J., Kärkkäinen, T.: The ONE simulator for DTN protocol evaluation. In: SIMUTools 2009: Proceedings of 2nd International Conference on Simulation Tools and Techniques, ICST, New York, NY, USA (2009)
8. Koubek, M., Rea, S., Pesch, D.: Event suppression for safety message dissemination in VANETs. In: 2010 IEEE 71st Vehicular Technology Conference (VTC 2010-Spring), pp. 1–5. IEEE (2010)
9. Lai, Y., Gao, X., Liao, M., Xie, J., Lin, Z., Zhang, H.: Data gathering and offloading in delay tolerant mobile networks. Wirel. Netw. 22(3), 959–973 (2016)
10. Lai, Y., Lin, Z.: Data gatherinzg in opportunistic wireless sensor networks. Int. J. Distrib. Sens. Netw. 8, 230198 (2012)
11. Lai, Y., Xie, J., Lin, Z., Wang, T., Liao, M.: Adaptive data gathering in mobile sensor networks using speedy mobile elements. Sensors 15(9), 23218 23248 (2015)

12. Lee, U., Magistretti, E., Gerla, M., Bellavista, P., Corradi, A.: Dissemination and harvesting of urban data using vehicular sensing platforms. IEEE Trans. Veh. Technol. **58**(2), 882–901 (2009)

13. Lee, U., Zhou, B., Gerla, M., Magistretti, E., Bellavista, P., Corradi, A.: Mobeyes: smart mobs for urban monitoring with a vehicular sensor network. IEEE Wirel. Commun. **13**(5), 52–57 (2006)

14. Lindgren, A., Doria, A., Schelén, O.: Probabilistic routing in intermittently connected networks. In: Service Assurance with Partial and Intermittent Resources, pp. 239–254 (2004)

15. Palazzi, C.E., Pezzoni, F., Ruiz, P.M.: Delay-bounded data gathering in urban vehicular sensor networks. Pervas. Mob. Comput. **8**(2), 180–193 (2012)

16. Paczek, B.: Selective data collection in vehicular networks for traffic control applications. Transp. Res. Part C: Emerg. Technol. **23**, 14–28 (2012). Data Management in Vehicular Networks. http://www.sciencedirect.com/science/article/pii/S0968090X1100180X

17. Zeng, J., Wang, T., Lai, Y., Liang, J., Chen, H.: Data delivery from WSNs to cloud based on a fog structure. In: 4th IEEE International Conference on Advanced Cloud and Big Data, no. 3, pp. 959–973 (2016, accepted)

Automatic Text Generation via Text Extraction Based on Submodular

Lisi Ai, Na Li, Jianbing Zheng, and Ming Gao(✉)

School of Data Science and Engineering,
East China Normal University, Shanghai 200062, China
irisinsh@163.com, nali0606@foxmail.com, zhengjb@js.chinamobile.com,
mgao@dase.ecnu.edu.cn

Abstract. Automatic text generation is the generation of natural language texts by computer. It has many applications, including automatic report generation, online promotion, etc. However, the problem is still a challenged task due to the lack of readability and coherence even there are many existing works studied it. In this paper, we propose a two-phase algorithm, which consists of text cleanup and text extraction, to automatically generate text from multiple texts. In the first phase, we generate paragraphs based on the topic modeling and clustering analysis. In the second phase, we model the text extraction as a set covering problem after we find the keywords in terms of the scores of TF-IDF, and solve the problem via employing the tool of submodular. We conduct a set of experiments to evaluate our proposed method and experimental results demonstrate the effectiveness of our proposed method by comparing with some comparable baselines.

Keywords: Automatic text generation · Massive information · K-Means · Submodular

1 Introduction

Automatic text generation is the generation of natural language texts by computer, which is a hot topic in both academia and industry. It can be applied to intelligent Question-Answer System, news report and online promotion, etc.

There are many state-of-art methods in this area and most of them already worked in the practice. Although great achievements have been made in this field, for example, the Newsblaster system [1] is a successful system from Columbia university, which is a news tracking tool to summarize the important news every day, it is still a challenged task due to the some problems, such as readability, coherence, a high-level complexity and so on.

We propose an automatic text generation algorithm, which consists of two phases: text cleanup and text extraction. Inspired by Lin [3], we employ the tool of submodular to solve the set covering problem. In summary, the main contributions of this paper are threefold:

© Springer International Publishing AG 2017
S. Song et al. (Eds.): APWeb-WAIM 2017 Workshops, LNCS 10612, pp. 237–246, 2017.
https://doi.org/10.1007/978-3-319-69781-9_23

- Based on topic model, we cluster documents into several clusters, which makes the generated text more readable and accurate.
- We model text extraction as a set covering problem which is solved by the tools of integer programming and submodular. Furthermore, we figure out the lower bound of performance of our proposed algorithm.
- To illustrate the promising results of our algorithm, we conduct extensive experiments on real data sets. Not only do experiments verify the feasibility of our method, but also it reveals our method outperforms the baselines.

Moreover, we describe the related work and preliminaries in Sect. 2. Section 3 and Sect. 4 show the problem formulation and algorithms. Experimental results are analysed in Sect. 5. Finally, we conclude the paper and discuss the future work in Sect. 6.

2   Related Work and Preliminaries

2.1   Related Work

Automatic text generation has been an important research field in the past few years. The methods of automatic text generation can be divided into four categories: text-to-text generation [10–17], meaning-to-text generation [5], data-to-text generation [6–8], image-to-text generation [9]. Text-to-text generation is the major research method and in this paper, we propose a text-to-text method to generate the new text.

Due to extracting the original sentences, text-to-text generation has a more stable semantic and grammatical structure. Li [10] proposed a method which generates new text by extracting sentences from original text, which has the information redundancy problem. Mihalcea and Tarau [16] proposed a sorting algorithm based on graphs, TextRank. They first build a graph associated with the text, where the graph vertices are representative for the units to be ranked. The sentences with high score are taken as the summary. In addition, submodular function is also used in text generation. Lin [3] proposed a method that given a set of objects $V = v_1, \ldots, v_n$ and a function $F : 2^V \rightarrow R$ which returns a real value for any subset $S \subseteq V$, this method is to find the subset of bounded size $|S| \leq k$ that maximizes the function, e.g., $argmax_{S \subseteq V} F(S)$. This method is similar to MMR [4], which compute $F(S)$ by sentence similarity. Although we also address the problem by submodular function, our approach is different from Lin [3]'s method since we compute $F(S)$ by set-covering ratio.

2.2   Preliminaries

A set function that satisfies the rule of diminishing marginal efficiency is called submodular function. For a set $V = \{v_1, v_2, v_3, \ldots, v_n\}$ and a function $f : 2^V \rightarrow R$. If $A \subseteq B \subseteq V$ and $e \in V - B$, then:

$$f(A \cup \{e\}) - f(A) \geq f(B \cup \{e\}) - f(B) \tag{1}$$

For a submodular function f and a limit condition C, we need to find the set S which satisfies the limit of C and can maximize the value of $f(S)$. The most basic greedy algorithm [18] for this NP hard is to add the largest incremental value which satisfies the conditional C at each iteration, so in i-th iteration:

$$S_i = S_{i-1} \cup \{argmax_e \Delta(e \mid S_{i-1})\} \tag{2}$$

In this equation, $\Delta(e \mid S_{i-1}) = f(S_{i-1} \cup \{e\}) - f(S_{i-1})$, where $f(S)$ satisfies the property of submodular. Obviously, a submodular function is monotonous and non-negative, i.e., for $\forall e \notin S, f(S \cup \{e\}) \geq f(S)$ or $\forall e \notin S, f(S \cup \{e\}) \leq f(S)$ and $\forall S \subseteq V, f(S) \geq 0$. The lower bound performance of this algorithm is 63.21% of optimal solution. From the property of submodular, we can define coverage function as weights of sentences so that automatic text generation satisfies the property of submodular and can be solved by this algortihm.

3   Problem Setup

3.1   Problem Formulation

Assume we have document cluster $D = \{d_1, d_2, \ldots, d_n\}$ and keyword set K. The goal is to find the sentence set $S = \{s_i \mid s_i \in d_j, d_j \in D\}$, which S can cover all keywords from K. Considering the applicability and effectiveness, we choose text extraction method. We model text extraction as a set covering problem, the definition as follows:

Definition 1 *(Text Extraction).* *Given keyword set $V = \{w_1, w_2, \ldots, w_{n1}\}$, the customized keyword set $U = \{u_1, u_2, \ldots, u_{n2}\}$ and sentence set $C = \{S_1, S_2, \ldots, S_m\}$. Assume that $S_j = \{w_k \mid w_k \in U \cup V\}$, the purpose is to find the minimal subset C' that $V \in U_{S \in C'} S$ and $\exists u_i \in U_{S \in C'} S$.*

3.2   Automatic Text Generation Process

The automatic text generation process is divided into two parts: corpus preprocessing, text generation.

1. Corpus Preprocessing.
   Because the high topic confusion of corpus, we classified it into different clusters. LDA (Latent Dirichlet Allocation) is a topic generation model proposed by Blei [15], which can generate the topic probability distribution of a document. In this paper, we use LDA model to predict distribution of topics in corpus and cluster documents by topic distribution. Finally, we obtain several document sets with lower confusion as candidate set of documents.
2. Text Generation.
   In order to generate a beautiful text structure, we classified the candidate set again. After clustering the candidate set into several clusters, we use each cluster to generate a paragraph.

According to the framework shown in Fig. 1, this approach contains two-phase: text cleanup and text extraction. In text cleanup phase, we obtain topic distribution of candidate documents by LDA model and use it to represent documents. Then, we cluster the documents into different topics. In text extraction phase, we use the scores of TF-IDF to define the keywords, we model the text extraction as a keyword set covering problem. Then, we solve the problem via employing the tool of submodular. The new text is composed of sentences extracted from each cluster.

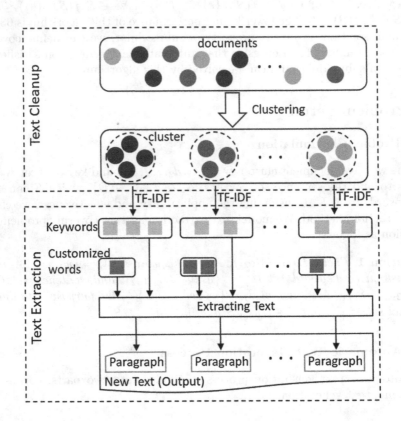

Fig. 1. The framework for automatic text generation

4   Algorithms

1. Keyword Set Generation
   TF-IDF (Term Frequency-Inverse Document Frequency) is a statistical method to evaluate the importance of a word in one document or one corpus. TF represents the frequency of one word appearing in the document. For one specified word t_i, the value of TF is denoted as $T_{i,j}$. IDF is a measure of the

importance of one word, for a specified word t_i, the IDF value is denoted as I_i. Then, we use variable $T - I_{i,j} = T_{i,j} \times I_i$ to obtain the value of TF-IDF. All in all, a high-frequency word has a larger TF-IDF value, so we can use this algorithm to obtain most important words and to filter out high-frequency non-keywords.

2. Text Extraction

After obtaining the set of keywords from the above algorithm and defining customized words, we model text extraction as a set covering problem, the formal description is as follows:

$$minimize |C'|, \text{ that is, } minimize \sum_{j=1}^{m} X_j$$

$$s.t. \ X_j \in \{0,1\}, j = 1, 2, \ldots, m$$

$$Y_{ij} \in \{0,1\}, i = 1, 2, \ldots, n1. j = 1, 2, \ldots, m$$

$$Z_{ij} \in \{0,1\}, i = 1, 2, \ldots, n2. j = 1, 2, \ldots, m$$

$$\sum_{j=1}^{m} X_j Y_{i,j} \geq 1, i = 1, 2, \ldots, n1. (1*)$$

$$\sum_{i=1}^{n2} \sum_{j=1}^{m} X_j Z_{i,j} \geq 1. (2*)$$

Among this, C' is a set of the extracted sentences, which composes the generated new text; $X_j = 1$ represents sentence S_j is contained in the sentence set C', $X_j = 0$ is opposite; $Y_{i,j} = 1$ means the i-th keyword w_i is in sentence S_j, $Y_{i,j} = 0$ is opposite; $Z_{i,j} = 1$ means the i-th customized keyword u_i is in sentence S_j, $Z_{i,j} = 0$ is opposite. Also, constraint (1*) is to ensure generated text covers all keywords; constraint (2*) is to ensure generated text at least covers one customized keyword.

To solve this set covering problem, one method is to employ the linear programming to approximate the integer programming, which is a NP-hard problem and we cannot theoretically figure out the lower bound of performance of it. Another method is using submodular (See Sect. 2.2) if we can define a set function satisfying the property of submodular. That is the focus of this paper.

In order to solve the set covering problem, we use the following objective function:

$$f(S) = \sum_{w_i \in \{w | w \in S\} \cap K / \{w \mid w \in C'\}} W_{w_i} + \sum_{w_j \in \{w \mid w \subset S\} \cap K_U / \{w \mid w \in C'\}} W_{w_j} \quad (3)$$

where w_i is a generated keyword, which occurs in sentence set S but not in extracted sentence set C', W_{w_i} is the weight value of keyword w_i. w_j is a customized keyword, which occurs in sentence set S but not in extracted sentence set C', W_{w_j} is the weight value of keyword w_j. Obviously, this function is a submodular. The algorithm is shown in Algorithm 1.

Algorithm 1. Text Extraction Algorithm with Submodular

Input:
    The set of sentences, V;
    The set of keywords, K;
    The set of customized keywords, K_U;
Output:
    The set of generated sentences, $C' \subseteq V$;
1: Initialize $C' = \emptyset$
2: **while** $(K! = \emptyset) or (C' \cap K_U = \emptyset)$ **do**
3:    **select** s, s.t. $s = arg \max\limits_{v \in V - C'} [f(C' \cup \{v\}) - f(C')]$
4:    $C' = C' \cup \{s\}$
5:    $\forall_{w \in s} K.remove(w)$
6:    $\forall_{u \in s} K_U.remove(u)$
7: **end while**
8: **return** C'

In the Algorithm 1, w is a word in the set of K, v is a word in the set of K_U. The statement of 2 row is the termination conditions of the algorithm, $K! = \emptyset$ means the keywords is not covered completely, $C' \cap K_U = \emptyset$ means that no customized keywords is covered, when satisfying both of them, program is terminated. After finding optimal sentence, we update K by removing keywords in this sentence and update K_U by removing customized words in this sentence. we employ hill-climbing algorithm to solve the set covering problem, so that we can use the least sentences to cover all keywords.

In this algorithm, we find unit s with the largest ratio of objective function gain to scaled cost. If adding s increases the objective function value and not violates the budget constraint, it is then selected and otherwise bypassed. In each iteration, we add the sentence with greatest weight into coverage set. The coverage set becomes the final output.

5   Experiments

5.1   Datasets and Parameters

The corpus which contains 1703 documents. After classifying those documents, we choose 4 classifications as candidate sets. The number of documents in each classification are 529, 245, 649 and 280 respectively. We pick 10 documents from each classification as experimental data and other documents are trained by LDA model to cluster by different topics.

Owning to no standard evaluation data set, we generate standard summarization by manual annotation. For each candidate set which contains 10 documents, we choose 3 volunteers to extract one summarization and there are 12 standard summarizations in all. In this paper, the default parameters of number of topics in LDA topic cluster, K value of K-Means and number of keywords in one classification are set as 50, 4 and 2 * #documents per cluster respectively.

5.2   Baselines

We compare our method with TextRank method based on graphs proposed by Mihalcea [16] and LinearPro method which approximates the set covering problem with a linear programming.

1. TextRank
   This algorithm builds graph by using sentences as nodes and similarities between sentences as weights of edges. TextRank algorithm is an unsupervised text generation extraction method with better results. Different methods have different computing method if sentence similarity.
2. LinearPro
   In Sect. 4, we have mentioned that two methods can solve the set covering problem, the one is integer programming which is a NP-hard problem. We can employ the linear programming to approximate the problem. However, we cannot figure out the lower bound of performance of this method, we use this method as baseline and we donate it as LinearPro. In this method, after getting the weights of each sentence, we need to define a threshold and sentence is extracted when the weight of the sentence is greater than the threshold, in this experiment, we set threshold as 0.8.

5.3   Evaluation Metrics

The method proposed in this paper is evaluated by Edmundson coincidence rate [17] and ROUGE evaluation standard [19].

- Edmundson
  The basic unit of Edmundson is sentence, the coincidence rate C is calculated by the following equation:

$$C = \frac{|S_g \cap S_s|}{|S_s|} \times 100\% \tag{4}$$

In this equation, S_g represents the set of extracted sentences and S_s represent standard text, generally, we use average value of coincidence rate (Ave) to reduce error.

- ROUGE
  In Edmundson method, we take sentences as compared objects, in this paper, we use ROUGE-N and ROUGE-S metrics. ROUGE-N can reflect the occurrence order of words, the equation is defined as:

$$ROUGE - N = \frac{\sum_{S \in \{ST\}} \sum_{ng \in S} C_M(ng)}{\sum_{S \in \{ST\}} \sum_{ngw \in S} C(ng)} \tag{5}$$

In this equation, ST represents standard text, $C_M(ng)$ represents the number of n-grams coexisted in new text and standard text. $C(ng)$ represents the occurrence number of n-grams in standard text.

5.4   Experimental Results Analysis

Based on the evaluation metrics of Edmundson coincidence rate, ROUGE1, ROUGE2, ROUGE S and ROUGE SU, we compare our algorithm with baselines.

Comparison in Edmundson Coincidence Rate. Table 1 is the comparison with TextRank, LinearPro and Submodular in Edmundson Coincidence Rate.

Table 1. Comparison with baselines in edmundson coincidence rate

Algorithm	Classification 1	Classification 2	Classification 3	Classification 4
TextRank(BM25)	0.0606	0.0	0.0192	0.0093
TextRank(Sim)	0.0303	0.0	0.0192	0.0093
LinearPro	0.0455	0.0213	0.0288	0.0093
Submodular	**0.09**	**0.0851**	**0.1346**	**0.0370**

From Table 1, we can find that for four classifications, Edmundson coincidence rate of our algorithm is higher a lot than other algorithms. For classification 2, TextRank(BM25) and TextRank(Sim) have not worked and the coincidence rate of LinearPro is only 0.0213, but the coincidence rate of our method is 0.0851, which is higher than LinearPro. Obviously, these experimental data indicates that our method outperforms than baselines.

Comparison in ROUGE. In the progress of experiment, we find the performance of TextRank(Sim) is better than TextRank(BM25), so in this part and next part, we use TextRank(Sim) and LinearPro to compare with Submodular algorithm in ROUGE-1, ROUGE-2, ROUGE-S, ROUGE-SU. The experimental results are shown in Fig. 2.

From Fig. 2, Submodular outperforms TextRank and LinearPro generally. Though LinearPro is better than Submodular in C1, but for C2, C3, C4, Submodular method have absolute superiority. Submodular is worse than TextRank(Sim) in R-2 for C3, but it is better than TextRank(Sim) in R-1, R-S and R-SU. From the perspective of sentence extraction, R-S and R-SU can reflect the sequence of sentences consistent with standard text more effectively and have more practical significance than R-2. Also, from the perspective of probability, our method has better results in most classifications. Therefore, our method is more effective than TextRank and LinearPro.

Comparison in Mean Value of ROUGE. Table 2 shows the mean values of R-1, R-2, R-S, R-SU of four classifications for TextRank(Sim), LinearPro and Submodular methods respectively. The mean value of Submodular in R-1, R-2, R-S, R-SU is higher than that of TextRank(Sim) and LinearPro algorithms.

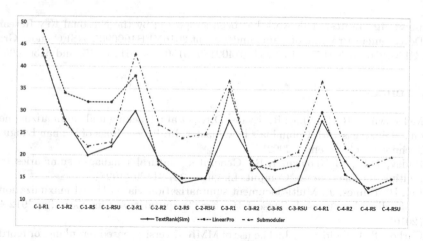

Fig. 2. Comparison with baselines in ROUGE

Thus, from the perspective of average effect, Submodular is superior to baselines. Even in the worst case, the average R-S value of Submodular is improved about 1.5 than LinearPro, the average R-2 value of Submodular is improved about 2 than TextRank(Sim).

Table 2. Comparison with baselines in mean value of ROUGE

Method	R-1	R-2	R-S	R-SU
TextRank(Sim)	32.25	21.25	14.50	16.25
LinearPro	37.75	21.50	19.25	20.00
Submodular	**40.25**	**23.25**	**20.75**	**22.25**

In summary, for those evaluation metrics, Submodular algorithm is more outstanding than other algorithms.

6   Conclusion and Future Work

In this paper, we propose a two-phase algorithm which consists of text cleanup and text extraction. We generate paragraphs based on the topic modeling and clustering analysis firstly. Then, we model the text extraction as a set covering problem after we find the keywords by TF-IDF. Experiments show the effectiveness of our proposed method by comparing with comparable baselines.

In the future work, we will improve the efficiency of our method. Also, we can consider about the position of sentences or words of title so that the generated text will be more consistent with standard summarization in the next work.

Acknowledgements. This work has been supported by the National Key Research and Development Program of China under grant 2016YFB1000905, NSFC under Grant Nos. U1401256, 61402177, 61672234, 61402180, 61502236, 61462017, and 61363005.

References

1. McKeown, K.R., Barzilay, R., Evans, D.K., et al.: Tracking and summarizing news on a daily basis with Columbia's newsblaster. In: Proceedings of Human Language Technology Conference (2002)
2. Radev, D.R., McKeovwn, K.R.: Generating natural languages summaries from multiple on-line sources. Comput. Linguist. **24**, 21–29 (1998)
3. Lin, H., Bilmes, J.: Multi-document summarization via budgeted maximization of submodular functions. In: Association for Computational Linguistics, pp. 912–920 (2010)
4. Carbonell, J., Goldstein, J.: The use of MMR, diversity-based reranking for reordering documents and producing summaries. In: International ACM SIGIR Conference on Research and Development in Information Retrieval, pp. 335–336 (1998)
5. Zhang, Y., Krieger, H.U.: Large-scale corpus-driven PCFG approximation of an HPSG. In: Association for Computational Linguistics, pp. 198–208 (2011)
6. Sripada, S., Reiter, E., Davy, I.: SumTime-Mousam: configurable marine weather forecast generator. Expert Update. **6**, 4–10 (2003)
7. Kukich, K.: Design of a knowledge-based report generator. In: Association for Computational Linguistics, pp. 145–150 (1983)
8. Portet, F., Reiter, E., Gatt, A., et al.: Automatic generation of textual summaries from neonatal intensive care data. Artif. Intell. **173**, 789–816 (2009)
9. Karpathy, A., Fei-Fei, L.: Deep visual-semantic alignments for generating image descriptions. In: Proceedings of IEEE Conference on Computer Vision and Pattern Recognition, pp. 3128–3137 (2015)
10. Li, S., Ouyang, Y., Wang, W., et al.: Multi-document summarization using support vector regression. In: Proceedings of DUC (2007)
11. Clarke, J., Lapata, M.: Global inference for sentence compression: an integer linear programming approach. J. Artif. Intell. Res. **31**, 399–429 (2008)
12. Filippova, K.: Multi-sentence compression: finding shortest paths in word graphs. In: Association for Computational Linguistics, pp. 322–330 (2010)
13. Thadani, K., McKeown, K.: Supervised Sentence Fusion with single-stage inference. In: IJCNL, pp. 1410–1418 (2013)
14. Fujita, A., Inui, K., Matsumoto, Y.: Exploiting lexical conceptual structure for paraphrase generation. In: Dale, R., Wong, K.-F., Su, J., Kwong, O.Y. (eds.) IJC-NLP 2005. LNCS, vol. 3651, pp. 908–919. Springer, Heidelberg (2005). doi:10.1007/11562214_79
15. Blei, D.M., Ng, A.Y., Jordan, M.I.: Latent Dirichlet allocation. J. Mach. Learn. Res. **3**, 993–1022 (2003)
16. Mihalcea, R., Tarau, P.: TextRank: bringing order into texts. In: Association for Computational Linguistics, pp. 404–411 (2004)
17. Edmundson, H.P.: New methods in automatic extracting. J. ACM (JACM) **16**, 264–285 (1969)
18. Nemhauser, G.L., Wolsey, L.A., Fisher, M.L.: An analysis of approximations for maximizing submodular set functions I. Math. Program. **14**, 265–294 (1978)
19. Lin, C.Y.: Rouge: a package for automatic evaluation of summaries. In: Text Summarization Branches Out: Proceedings of ACL-2004 Workshop, pp. 74–81 (2004)

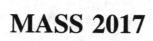

MASS 2017

Collaborative Heterogeneous Information Embedding for Recommender Systems

Zhen Lv[1], Haixia Zhang[1], Dalei Wu[2],
Chuanting Zhang[1], and Dongfeng Yuan[1(✉)]

[1] Shandong Provincial Key Laboratory of Wireless Communication Technologies,
Shandong University, Jinan 250100, Shandong, China
zhenlv_2011@163.com, chuanting.zhang@gmail.com,
{haixia.zhang,dfyuan}@sdu.edu.cn
[2] The University of Tennessee at Chattanooga, Chattanooga, TN 37403, USA
dalei-wu@utc.edu

Abstract. Collaborative Filtering (CF) is one of the most popular frameworks for recommender systems. However, sparsity of user-item interactions degrades the performance of CF significantly. Using auxiliary information is a common way to solve this sparsity problem. Heterogeneous information networks (HINs), which contains a plurality of types of nodes or rich relations between nodes, make it promising to boost the performance of recommendations. In this paper, by integrating a rich variety of heterogeneous information of items into CF, we propose a novel hybrid recommendation method called Collaborative Heterogeneous Information Embedding (CHIE). CHIE jointly performs fused representation learning for items in HIN and Probabilistic Matrix Factorization (PMF), a model-based CF, for the ratings matrix. Moreover, We conduct experiments on a real movie recommendation network, which show that our approach outperforms the state-of-the-art recommendation techniques.

Keywords: Recommender systems · Heterogeneous information · Probabilistic matrix factorization

1 Introduction

Due to the explosive growth of information, people are highly distressed about information overload and have trouble making choices every day. Recommendation is one of the most effective information filtering and discovery techniques. Among different recommendation technologies, CF-based methods [13] making use of historical interactions or preferences like ratings by users, is one of the most ubiquitously used frameworks for recommender systems. However, conventional CF-based methods suffer from data sparsity a lot, since the interactions between users and items are consistent with power law distributions approximately, i.e., the majority of users rated a few items and there is a great deal of empty values

S. Song et al. (Eds.): APWeb-WAIM 2017 Workshops, LNCS 10612, pp. 249–256, 2017.
https://doi.org/10.1007/978-3-319-69781-9_24

in interaction matrix. It is particularly gratifying that recommendation performance of CF can be improved further by using auxiliary information [12], which is also the most general way to alleviate sparsity problem.

Extensive algorithms have been proposed that use auxiliary information to boost the accuracy of CF. For example, TrustSVD [2] combines trust relationship of users in social networks with the rating data. CTR [16] adds textual information of items such as comments and descriptions into CF. However, the methods above only utilize single type of information of users or items. Recently, Heterogeneous Information Network (HIN) [14], which contains multiple types of entities or relations, can lead to better recommendation performance [18]. Hence, the study of recommendations on HIN begin to draw much attention. [19] proposed Hete-MF by combining user ratings with multiple entity similarity metrics. Jamali and Lakshmanan proposed HeteroMF [3] to integrate a general latent factor and context-dependent latent factors. On the other hand, some attempts of making recommendations by the means of deep learning models have been tried. For example, [8] introduced neural network to Recommender system (RS) for the first time, AutoRec [9] made rating prediction using autoencoder to make the output of the model the same as the input. Recently a novel hierarchical Bayesian deep learning model called CDL [17] has been proposed. CDL bridges the gap between Stacked Denoising Autoencoders (SDAE) [15], which aims at learning the latent features of textual auxiliary information, and a model-based CF method [7] for the ratings matrix. However due to the inherent limitation of the bag of words model, CDL can not understand textual information effectively.

In this paper, we propose a novel hybrid recommendation method named CHIE, which jointly learn the latent vectors in CF as well as item representations based on different relations between entities in HIN. The proposed method captures different semantic information by fusing different item representations obtained by Bayesian deep learning framework according to different meta paths [14] and experimental results show that our method performs better than the state-of-the-art algorithms.

The main contributions of this paper are as follows.

- We propose a novel hybrid recommendation method named CHIE by combining deep learning algorithm, which can capture representations of items in HIN, with PMF for recommendation.
- To take advantage of the heterogeneity in HIN, we capture different semantic information by fusing the rating prediction results obtained by Bayesian deep learning framework according to different meta paths.
- Empirical studies in extended MovieLens datasets demonstrate the advancement of our methodology.

The remainder of this paper is structured as follows. In Sect. 2, we introduce the preliminary about the research in this paper. And in Sect. 3, we describe the detail of the proposed method. Next, in Sect. 4, we illustrate the experimental results. Finally, we conclude the work in Sect. 5.

2   Heterogeneous Information Network

An information network is called HIN when it contains more than one types of nodes or relations [11]. Figure 1a demonstrates a tiny network instance on extended MovieLens dataset, which is a typical HIN. The network schema [14] for an information network is denoted as $S = (A, R)$, where A is the entity type set and R is the relation type set. Network schema specifies type constraints and tells how many types of objects there are in the network and where the possible links exist. Figure 1b illustrates network schema of the HIN in Fig. 1a. In this example, it contains seven types of nodes and twelve different link types. Different from homogeneous networks, two objects in a HIN can be connected via different paths with different physical meanings. These paths can be categorized as meta paths.

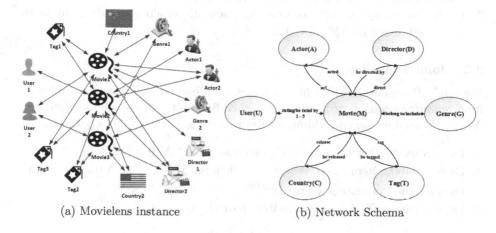

(a) Movielens instance   (b) Network Schema

Fig. 1. Movielens instance and network schema.

A meta path [14] P is a path defined on a schema $S = (A, R)$, and is denoted in the form of $A_1 \xrightarrow{R_1} A_2 \cdots \xrightarrow{R_l} A_{l+1}$, which can be abbreviated as $A_1 A_2 \cdots A_{l+1}$ if there is no ambiguity. Different meta paths represent different semantics: two movies are similar to each other might because they share the same actor (along meta path MAM), or another possibility is that they belong to the same genre (along meta path MGM). Taking the semantic information into consideration will lead to more subtle knowledge discovery and recommendation.

3   Proposed Method

3.1   Item Representation

Considering that different meta path indicates different meaning, we derive item representation according to K meta paths, which can make full use of the rich semantic information contained in HIN. Note that K is determined by the corresponding network schema. The process can be reformulated with more detail as follows.

1. For MovieLens data in Fig. 1a, we extract six meta paths from its network schema shown as Fig. 1b: $P_1 = $ MTM, $P_2 = $ MAM, $P_3 = $ MCM, $P_4 = $ MDM, $P_5 = $ MUM and $P_6 = $ MGM. Note that there are infinite meta paths in HIN if there is no restrictions, nevertheless, [14] shows that a very long meta path may be misleading, therefore, in this paper, we only consider meta paths that is relatively short as well as containing a clear and concise semantic meaning.
2. By exploring the tiny heterogeneous MovieLens instance shown in Fig. 1a, we get M-T relationship matrix along meta path MTM. Besides, similarity between two movies along MTM is modeled as cosine similarity in M-T matrix. In this matrix, we assume each row vector in n dimension space represents a movie, and n is the number of movies.
3. In the same way, we get six item representations along six different meta paths, expressed as $X_c^{(k)}$, $k = 1, 2 \cdots 6$.

3.2 Joint Learning

We apply Bayesian SDAE [1,5] to extract item embeddings based on different item representations respectively. The embedding vector in the middle layer, i.e., $X_{L/2,j*}$ is used as the embedding for item entity j. For each meta path P_k,

1. Draw a user latent vector of each user i as $u_i^{(k)} \sim \mathcal{N}(0, \lambda_u^{-1}I)$.
2. Draw a latent item offset vector of each item j as $e_j^{(k)} \sim \mathcal{N}(0, \lambda_v^{-1}I)$, and then set the item latent vector as $v_j^{(k)} = e_j^{(k)} + X_{L/2,j*}^{(k)}$.
3. Draw a rating R_{ij} for each user-item pair (i, j):

$$R_{ij} \sim \mathcal{N}((u_i^{(k)})^T v_j^{(k)}, C_{ij}^{-1}), \tag{1}$$

where C_{ij} is a confidence parameter that $C_{ij} = a$ if $R_{ij} = 0$ and $C_{ij} = b$ otherwise.

3.3 Learning the Parameters

It is intractable to Compute the full posterior of the parameters. Maximizing the posterior probability of u, v, W and b is equivalent to maximizing the log-likelihood [16] as follows:

$$\mathcal{L} = -\frac{\lambda_u}{2} \sum_i \| u_i \|_2^2 - \frac{\lambda_v}{2} \sum_j \| v_j - X_{\frac{L}{2},j*}^T \|_2^2 - \frac{\lambda_w}{2} \sum_l (\| W_l \|_F^2 + \| b_l \|_2^2)$$

$$-\frac{\lambda_n}{2} \sum_j \| X_{L,j*} - X_{c,j*} \|_2^2 - \frac{\lambda_s}{2} \sum_l \sum_j \| \sigma (X_{L-1,j*}W_l + b_l) - X_{L,j*} \|_2^2$$

$$-\sum_{i,j} \frac{C_{ij}}{2} (R_{ij} - u_i^T v_j)^2. \tag{2}$$

We adopt coordinate descent algorithm similar to [17], which iteratively optimizes a latent variable while fixing the remaining variables. Note that we get different $u_i^{(k)}$ and different $v_j^{(k)}$ based on different $X_{c,j*}^{(k)}$.

3.4 Results Merging

After obtaining the user and item low-rank representation pair $(u_i^{(k)}, v_j^{(k)})$, we define a recommendation model as follows:

$$\widehat{R}_{i,j} = \sum_{k=1}^{K} \theta_k \cdot \left(u_i^{(k)}\right)^T v_j^{(k)}, \qquad (3)$$

where θ_k is the weight for user and item low-rank representation pair based on the k-th meta path, and we can get θ_k easily using a supervised machine learning method.

With the recommendation model in Eq. 3, given a user, we can now predict the recommendation scores to all items, so that we can recommend items to the user according to the predicted scores.

4 Experimental Results and Analysis

4.1 Data Sets and Evaluation Methodology

The dataset contains 2113 users and 10197 movies, published by GroupLens research group[1], and we would like to call it extended MovieLens dataset in this paper. The dataset is shown in Table 1. We define the density of a dataset as the existing ratings in U-I matrix account for the ratio of all possible ratings, in that case the density of extended MovieLens dataset is 3.97%. It's obvious that conventional CF-based methods using U-I matrix only suffer from data sparsity a lot.

Table 1. Detailed information about extended MovieLens dataset.

Types	User	Movie	Actor	Director	Country	Genre	Tag
Notation	U	M	A	D	C	G	T
Numbers	2113	10197	21185	4060	72	20	13222

We select two most classical metrics in rating prediction task: Mean Absolute Error (MAE) and Root-Mean-Square Error (RMSE). The larger the values of MAE and RMSE are, the lower the prediction effectiveness is.

---

[1] http://www.grouplens.org.

4.2   Performance Comparison and Analysis

In this section, we directly set $a = 0.01$, $b = 1$, $N = 50$ and perform grid search on the hyperparameters λ_u, λ_v, λ_n, λ_s, and λ_w. For the grid search, we split the training data and use 5-fold cross validation. In order to show the performance improvement of our CHIE algorithm, we compare our algorithm with some state-of-the-art algorithms in extended MovieLens dataset and the results are shown in Figs. 2 and 3. The method Userave is the mean value of the rating rated by the target user and the method Itemave is the mean value of rating to the target item. We implemented PMF [7] and SVD++ [4] as two MF-based baselines. As to the deep learning method, we also utilized AutoRec [9] to compare with the proposed method. In addition, we realised two heterogeneous neighborhood-based CF algorithms using the same meta paths as the proposed method, in which PathSim [14] similarity and HeteSim [10] similarity are utilized.

Figure 2 presents the performances in MAE of all methods aforementioned. SVD++, PathSim-CF and HeteSim-CF gain much better performance than Userave, Itemave and AutoRec. This phenomenon verifies our discussion that

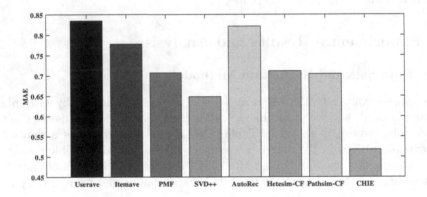

Fig. 2. Performance comparison in MAE.

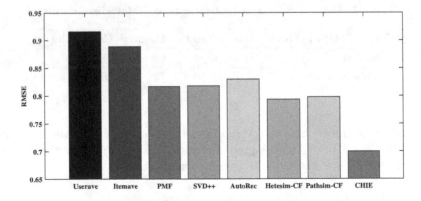

Fig. 3. Performance comparison in RMSE.

using auxiliary information can improve recommendation performance, for example, SVD++ introduces historical implicit feedback information and PathSim-CF adds heterogeneous information. However, our proposed method CHIE always outperforms other algorithms. In terms of RMSE, as shown in Fig. 3, Userave and Itemave are the two worst-performing methods since they only perform a simple and fix average calculating operation for the ratings. As an efficient algorithm, PMF outperforms Userave and Itemave by about 10%. Besides, SVD++ makes a more accurate recommendation prediction. The method AutoRec has a low accuracy possibly because it only uses user-item interaction information and also suffers from sparsity problem. As far as PathSim-CF and HeteSim-CF are concerned, they yield better performance in RMSE than other algorithms except for our proposed method because these two methods also utilize rich variety of information in HIN. Our proposed recommendation model beats all baseline methods in the experimental dataset in MAE and RMSE. The reason is that CHIE takes good advantage of various information in HIN and captures different semantic meanings along different meta paths. Moreover, it leverages excellent representation ability of deep learning as well.

5 Conclusions

In this paper, we studied recommendation in the scope of HIN and proposed a novel hybrid recommendation method called CHIE. Firstly, we capture different item representations with different semantic meaning based on various meta paths in HIN. Next we jointly learn latent features of users and items using a Bayesian deep learning model, which combines SDAE and PMF. Finally, we compared the proposed approach to several widely employed recommendation techniques. Empirical studies show that our method can improve the recommendation quality tremendously. Mobility analytics from spatial and social data is one of the hot emerging topics [6], in the future, we plan to explore more effective representation approaches by considering the time and space factors of user interests to further improve the performance of recommendation.

Acknowledgments. The work presented in this paper was supported in part by the Special Project for Independent Innovation and Achievement Transformation of Shandong Province (2013ZHZX2C0102, 2014ZZCX03401).

References

1. Bengio, Y., Yao, L., Alain, G., Vincent, P.: Generalized denoising auto-encoders as generative models. In: Proceedings of the 26th International Conference on Neural Information Processing Systems, NIPS 2013, pp. 899–907. Curran Associates Inc., USA (2013)
2. Guo, G., Zhang, J., Yorke-Smith, N.: Trustsvd: Collaborative filtering with both the explicit and implicit influence of user trust and of item ratings. In: Proceedings of the Twenty-Ninth AAAI Conference on Artificial Intelligence, AAAI 2015, pp. 123–129. AAAI Press (2015)

3. Jamali, M., Lakshmanan, L.: HeteroMF: recommendation in heterogeneous information networks using context dependent factor models. In: Proceedings of the 22nd International Conference on World Wide Web, WWW 2013, pp. 643–654. ACM, New York (2013)
4. Koren, Y.: Factorization meets the neighborhood: a multifaceted collaborative filtering model. In: Proceedings of the 14th ACM SIGKDD International Conference on Knowledge Discovery and Data Mining, KDD 2008, pp. 426–434. ACM, New York (2008)
5. MacKay, D.J.: A practical bayesian framework for backpropagation networks. Neural Comput. 4(3), 448–472 (1992)
6. Qu, Q., Chen, C., Jensen, C.S., Skovsgaard, A.: Space-time aware behavioral topic modeling for microblog posts. IEEE Data Eng. Bull. 38(2), 58–67 (2015)
7. Salakhutdinov, R., Mnih, A.: Probabilistic matrix factorization. In: International Conference on Neural Information Processing Systems, pp. 1257–1264 (2007)
8. Salakhutdinov, R., Mnih, A., Hinton, G.: Restricted Boltzmann machines for collaborative filtering. In: Proceedings of the 24th International Conference on Machine Learning, ICML 2007, pp. 791–798. ACM, New York (2007)
9. Sedhain, S., Menon, A.K., Sanner, S., Xie, L.: Autorec: autoencoders meet collaborative filtering. In: Proceedings of the 24th International Conference on World Wide Web, WWW 2015 Companion, pp. 111–112. ACM, New York (2015)
10. Shi, C., Kong, X., Huang, Y., Philip, S.Y., Wu, B.: Hetesim: a general framework for relevance measure in heterogeneous networks. IEEE Trans. Knowl. Data Eng. 26(10), 2479–2492 (2014)
11. Shi, C., Li, Y., Zhang, J., Sun, Y., Philip, S.Y.: A survey of heterogeneous information network analysis. IEEE Trans. Knowl. Data Eng. 29(1), 17–37 (2017)
12. Shi, Y., Larson, M., Hanjalic, A.: Collaborative filtering beyond the user-item matrix: a survey of the state of the art and future challenges. ACM Comput. Surv. 47(1), 3:1–3:45 (2014)
13. Su, X., Khoshgoftaar, T.M.: A survey of collaborative filtering techniques. Adv. Artif. Intell. 2009, 4:2 (2009)
14. Sun, Y., Han, J., Yan, X., Yu, P.S., Wu, T.: Pathsim: meta path-based top-k similarity search in heterogeneous information networks. Proc. VLDB Endow. 4(11), 992–1003 (2011)
15. Vincent, P., Larochelle, H., Lajoie, I., Bengio, Y., Manzagol, P.A.: Stacked denoising autoencoders: learning useful representations in a deep network with a local denoising criterion. J. Mach. Learn. Res. 11(Dec), 3371–3408 (2010)
16. Wang, C., Blei, D.M.: Collaborative topic modeling for recommending scientific articles. In: Proceedings of the 17th ACM SIGKDD International Conference on Knowledge Discovery and Data Mining, KDD 2011, pp. 448–456. ACM, New York (2011)
17. Wang, H., Wang, N., Yeung, D.Y.: Collaborative deep learning for recommender systems. In: Proceedings of the 21th ACM SIGKDD International Conference on Knowledge Discovery and Data Mining, KDD 2015, pp. 1235–1244. ACM, New York (2015)
18. Yang, B., Lei, Y., Liu, D., Liu, J.: Social collaborative filtering by trust. In: Proceedings of the Twenty-Third International Joint Conference on Artificial Intelligence, IJCAI 2013, pp. 2747–2753. AAAI Press (2013)
19. Yu, X., Ren, X., Gu, Q., Sun, Y., Han, J.: Collaborative filtering with entity similarity regularization in heterogeneous information networks. IJCAI HINA 27 (2013)

Comparative Research for Social Recommendations on VK

Rustam Tukhvatov and Joo Young Lee[(✉)]

Innopolis University, Innopolis 420500, Russia
{r.tukhvatov,j.lee}@innopolis.ru

Abstract. Recommender system is one of the most important component for many companies and social networks such as Facebook and YouTube. A recommendation system consists of algorithms which allow to predict and recommend friends or products. This paper studies to facilitate finding like-minded people with same interests in social networks. In our research we used real data from the most popular social network in Russia, VK (Vkontakte). The result shows that majority of users in VK tend not to add possible users with whom they have common acquaintances. We also propose a topology based similarity measure to predict future friends. Then we compare our results with the results of other well known methods and discuss differences.

Keywords: Recommender system · Graph theory · Link prediction · Similarity

1 Introduction

Social networks help people to interact through Internet. Nowadays social network services allow to share interests via texts, music, video, etc. Clearly, people want to get more information and new contacts as fast as possible and the information should be relevant to their preferences.

Link recommendation has become one of the most important features in online social networks and has been an active research area [2,6–8,12,14]. There are well known examples of link recommendation such as "People You May Want to Hire" on LinkedIn, "You May Know" on Google+ and "People You May Know" on Facebook. Given the tremendous academic and practical interests in link recommendation [4], we examined existing approaches and applied recommendation engine for VK social network.

A social network is a graph as a data structure, the users are nodes and the users' friendships (relations) are edges [9] We aim to understand factors which may affect the emergence of new edges and try to predict future connections in social networks. In Sect. 2 we introduce some of major existing methods on link prediction. In Sect. 3, we present existing methods which we test and compare our results with on VK datasets. In Sect. 6, we show our results and discuss implications. Finally, in Sect. 7, we conclude our research.

© Springer International Publishing AG 2017
S. Song et al. (Eds.): APWeb-WAIM 2017 Workshops, LNCS 10612, pp. 257–266, 2017.
https://doi.org/10.1007/978-3-319-69781-9_25

2   Related Work

[15] studies understanding users' temporal behaviors in social network platforms. The evolution of social network services is driven by the interplay between users' preferences and social network structures. Authors argue that users' future preference behavior is affected by the network around them and the homophily effect. In [1] authors demonstrate the developed Social Poisson Factorization (SPF), a probabilistic model that incorporates social network information into a traditional factorization method. SPF introduces the social aspect of algorithmic recommendation. They develop a scalable algorithm for analyzing data with SPF, and demonstrate that it outperforms competing methods on six real-world datasets.

[16] studied the social influence problem in a large microblogging network, Weibo.com.[1] They investigate (re)tweet behaviors of users by considering ego network of users. In our proposed method, we expand another step by considering second level degrees of users. [3] demonstrates online consumers' behavior and try to explain ways to improve ad (advertising) targeting systems. Researchers use information in emails such as purchase logs and communications between users to find patterns and their interaction. For example, authors measured the effect of gender and age which showed that a female email user is more likely to be an online shopper than an average male email user. Also the spending ability goes up with the age until age of 30, then stabilizes in the early 60s, and starts to drop afterward. Such findings help to predict future customers' behavior and make purchases more pleasant for consumers.

Measuring similarity between nodes is the main task in link prediction problem. In [13], authors constructed a new way to measure similarity between nodes based on game-theoretic interaction index. The basic form of the interaction index is built upon two solution concepts from game theory: the Shapley value and the Banzhaf index. It is also generalized to a wider class of solution concepts: Shapley value, Banzhaf index. Authors showed that using their approach, it is possible to improve existing results in link prediction and community detection problems.

3   Methods

The link prediction problem is connected with network structure. In our research we use similarity between users using information from their second level degrees.

3.1   Problem Statement

Given a snapshot of a social network $G^t = (V^t, E^t)$ where V is a set of users and E is a set of relationships, we aim to find a relationship $(v_i, v_j) \notin E^t$ and $(v_i, v_j) \in E^{t'}$ where $t < t'$.

---

[1] The most popular Chinese microblogging service.

The link prediction problem is the problem of predicting future connections in a network which may form in the network. Here we briefly list some measures which we examine in this study.

$$CosineSimilarity(A, B) = cos(\theta) = \frac{f(A)f(B)}{|f(A)||f(B)|} \tag{1}$$

$$CommonNeighbors(A, B) = f(A) \cap f(b) \tag{2}$$

$$JaccardSimilarity(A, B) = \frac{|f(A) \cap f(B)|}{|f(A) \cup f(B)|} \tag{3}$$

$$AdamicAdarIndex(A, B) = \sum_{w \in f(A) \cap f(B)} \frac{1}{\log |f(w)|} \tag{4}$$

where $(f(w))$ denotes the set of neighbors of w.

3.2   Second Common Neighbor Similarity

We propose a new method that measures similarity between users using second degrees [5, 10].

$$SecondCommonNeighbors(A, B) = | \bigcup_{i \in f(A)} f(i) \cap \bigcup_{j \in f(B)} f(j)| \tag{5}$$

$$\text{where } f() \text{ is user's friends,}$$
$$f(A) \text{ friend of user } A,$$
$$f(i) \text{ is the set of neighbors of } i.$$

We propose another simple metric to compare with the proposed similarity measure which is based on the shortest distance between users.

$$ShortestPathIndex(A, B) = \text{Length Of Shortest Path Between } A \text{ and } B \tag{6}$$

4   Experiments

4.1   Collection of Dataset

We collected several snapshots of networks from VK to test link prediction methods. In order to make comparisons easier, we preprocess the data so that the resulting networks contains the same users since newly joined users in latter networks are not predictable. In the next step, we exclude users who are suspected to be not regular individuals such as groups, commercial accounts, and celebrities. We removed users with more than 500 friends based on the mean number of friends in VK which is 240. Datasets contain users' profile information, friends and posts from their walls as shown in Fig. 1.

	id	personal	friends	wallPosts
0	2064386.0	[{'uid': 2064386, 'country': 1, 'counters': {'...	[2772.0, 4987.0, 8680.0, 11380.0, 11698.0, 132...	[111199386, 'Левитанский Ю.Д. Дети <b...
1	63802027.0	[{'home_town': 'Москва', 'movies': '', 'about'...	[701.0, 77777.0, 88484.0, 196957.0, 204655.0, ...	[0, 'Ура! С днем варенья, дружище! И пускай вс...
2	241284438.0	[{'home_town': '', 'movies': '', 'photo_50': '...	[36417.0, 84922.0, 125032.0, 146957.0, 211166...	[8095605, '', 4698690, '', 4247839, '', 177117...
3	7.0	[{'uid': 7, 'counters': {'followers': 5902, 'v...	[6.0, 9.0, 10.0, 14.0, 17.0, 21.0, 22.0, 23.0,...	[]
4	262153.0	[{'uid': 262153, 'country': 1, 'counters': {'u...	[651.0, 1174.0, 3775.0, 5068.0, 7720.0, 10437....	[0, '', -32179668, 'Новогоднее печенье Рече...

Fig. 1. Extracted data from VK.

Table 1. VK networks with different time stamps.

Date	Number of nodes	Number of edges	Average degree
23.12.2016	942469	1475811	3.1318
24.01.2017	950503	1490873	3.1370
23.02.2017	936847	1465214	3.1280
24.03.2017	939985	1469854	3.1274

Table 2. VK networks after preprocessing.

Date	Number of nodes	Number of edges	Average degree	Excess nodes
23.12.2016	190550	712014	7.4733	7412
24.01.2017	190550	716478	7.5201	5459
23.02.2017	190550	712196	7.4752	1083
24.03.2017	190550	713030	7.4839	340

We use VK networks with 4 different timestamps as summarized in Table 1.

We further processed the data by removing nodes whose degree is less than two and newly joined nodes that did not exist in the starting graph. The processed networks are summarized in Table 2.

To evaluate the quality of predictions we use commonly accepted measures: *precision, recall, accuracy* and *F1 score*.

5   Evaluations

To have a base line we started from the most commonly used measures of similarity [11]: Cosine 1, Common neighbor 2, Jaccard 3 and Adamic-Adar 4.

To examine popular similarity measures to predict links, we used data from Vkontakte (VK) social network. First, we processed Facebook dataset. The average number of users' friends is 23 and for VK is 246 and to efficiently process the big data we have reduced from 10728 nodes (2332068 edges) to 4000 user, so that it contains 4000 nodes and 1122226 edges. To evaluate the performance we deleted 30% edges and then try to predict the deleted ones. After deleting the number of edges is 784499. We used 4 similarity measures: Common neighbor, Cosine, Jaccard and Adamic-Adar.

The threshold value is chosen as when we test Facebook dataset: 0.6 for cosine and 0 for others similarity measures. Table 3 illustrates the most popular metrics to evaluating of similarity measures: Precision, Recall, Accuracy and f1-score.

Table 3. Evaluating measures table for VK (baseline).

	Common neighbor	Cosine	Jaccard	Adamic-Adar
Precision	0.002936	**0.019121**	0.002936	0.014459
Recall	0.008744	**0.171432**	0.008744	0.005208
Accuracy	0.837793	**0.959589**	0.837793	0.822696
F1 score	0.004395	0.0005134	0.004395	**0.007658**

The quality of predictions is lower than for Facebook dataset. One of the reasons that we suspect is that the graph (for VK) is very sparse and the relationships are concentrated in small regions. The results of the VK dataset shows that Adamic-Adar similatity index the best evaluating measure according to F1-score and Cosine for others.

6 Results

The dataset contains 13951 users, after excluding users who have more than 500 friends. To make predictions, we deleted 30% of the edges.

The Table 4 illustrates evaluations of similarity measures: Precision, Recall, Accuracy and f1-score. We found that Jaccard similarity has the best values in Precision and F1-score.

Table 4. Evaluating measures table for VK (deleting approach).

	Cosine	Jaccard	Adamic-Adar
Precision	0.0004111	0.0053608	0.0289651
Recall	0.0428219	0.0403065	0.0016984
Accuracy	0.0085148	0.9190919	0.8309117
F1 score	0.0008144	**0.0094629**	0.0032087

Real Data. We compare VK dataset with 6966 users which was collected on the 22nd of December (2016) with 1506714 edges and the dataset with the same users which was collected on the 24th of January (2017) which contains 1523040 edges. The Table 5 illustrates the obtained results.

Table 5. Evaluating measures table for VK (real data comparisons).

	Cosine	Jaccard	Adamic-Adar
Precision	0.0000038	0.0000297	0.0001549
Recall	0.0001232	0.0001102	0.0000779
Accuracy	0.0030727	0.8596805	0.9532686
F1 score	0.0000073	0.0000468	0.0001037

6.1 Evaluation of Other Approaches

Tables 6, 7 and 8 illustrate results of the similarity measures when we limited the number of predictions by selecting the top 2% from the best similarity indexes for 1000 randomly selected users.

We omit Common Neighbor similarity here since it shows similar results with Jaccard index and Cosine. Instead we add two other measures, i.e., Second Neighbor similarity in Eq. (5) and Shortest Path Length in Eq. (6). The main outcome of the results is that TP is equal zero which implies that we predict very small number of potential relationships.

Table 6. Illustration of various rates (data from the January)

	jaccard
tp	0
fn	534
fp	1006

(a) Jaccard rates.

	adamic
tp	0
fn	534
fp	1006

(b) Adamic-adar rates.

	second
tp	0
fn	4608
fp	1006

(c) Second Neighbor rates.

	second
tp	0
fn	4800
fp	1006

(d) Shortest path rates.

Top 25. In this experiment we select 25% from the best similarity indexes for 1000 randomly selected users. Results are summarized in Tables 9, 10 and 11. The values for TP has increased but we still have big false predictions.

6.2 Comparison of Second Neighbor Approach

In this section, we have used facebook dataset which is previously presented. We have deleted 30 % edges, then make prediction for 1000 randomly selected users. Table 12 illustrates that the proposed second neighbor measure shows the best TP and lowest FP with smaller FN compared to the shortest path measure.

Table 7. Illustration of various rates (data from February 2017).

	jaccard
tp	0
fn	534
fp	1252

(a) Jaccard rates.

	adamic
tp	0
fn	534
fp	1252

(b) Adamic-adar rates.

	second
tp	0
fn	4608
fp	1252

(c) Second Neighbor rates.

	shortest
tp	0
fn	4800
fp	1252

(d) Shortest path rates.

Table 8. Illustration of various rates (data from March 2017).

	jaccard
tp	0
fn	534
fp	1567

(a) Jaccard rates.

	adamic
tp	0
fn	534
fp	1567

(b) Adamic-adar rates.

	second
tp	0
fn	4608
fp	1567

(c) Second Neighbor rates.

	shortest
tp	0
fn	4800
fp	1567

(d) Shortest path rates.

Table 9. Illustration of various rates (data from January 2017).

	jaccard
tp	2
fn	24336
fp	397

(a) Jaccard rates.

	adamic
tp	2
fn	24336
fp	397

(b) Adamic-adar rates.

	second
tp	2
fn	211130
fp	397

(c) Second Neighbor rates.

	shortest
tp	2
fn	219836
fp	397

(d) Shortest path rates.

Table 10. Illustration of various rates (data from February 2017).

	jaccard
tp	3
fn	24335
fp	459

(a) Jaccard rates.

	adamic
tp	3
fn	24335
fp	459

(b) Adamic-adar rates.

	second
tp	3
fn	211129
fp	459

(c) Second Neighbor rates.

	shortest
tp	3
fn	219835
fp	459

(d) Shortest path rates.

Table 11. Illustration of various rates for 25% top predictions (data from March 2017).

	jaccard
tp	4
fn	24334
fp	535

(a) Jaccard rates.

	adamic
tp	4
fn	24334
fp	535

(b) Adamic-adar rates.

	second
tp	4
fn	211128
fp	535

(c) Second Neighbor rates.

	shortest
tp	4
fn	219834
fp	535

(d) Shortest path rates.

Table 12. Illustration of various rates (data from the Facebook dataset).

	jaccard
tp	1776
fn	105520
fp	10212

(a) Jaccard rates.

	adamic
tp	1776
fn	105520
fp	10212

(b) Adamic-adar rates.

	second
tp	**1791**
fn	545865
fp	**10197**

(c) Second neighbor rates.

	shortest
tp	1791
fn	922919
fp	10197

(d) Shortest path rates.

7   Conclusions

There exist many different types of link prediction approaches and they can be roughly categorized into learning-based methods and proximity-based methods. In our research we considered proximity-based methods, namely, cosine similarity, common neighbor, jaccard similarity, adamic-adar index, second neighbor and shortest path. All of them are structural proximity-based methods which does not consider nodal information. We have thoroughly tested the measures on different datasets. Many existing studies make prediction only for randomly deleted edges due to lack of real datasets. We collected our own snapshots of graphs from VK social network within four months. To facilitate the collected dataset for prediction, we filtered users that have too many links and also maintain the same users for each snapshot since we predict links but not new users in this study.

The results do not conclude which approach is superior in most of the cases. The prediction method to use for real networks should be chosen with heuristics given the nature of the network. In this study, we showed that there is a great difference with the same measures in VK and facebook datasets.

References

1. Chaney, A.J.-B., Blei, D.M., Eliassi-Rad, T.: A probabilistic model for using social networks in personalized item recommendation. In: Proceedings of the 9th ACM Conference on Recommender Systems, RecSys 2015, Vienna, Austria, 16–20 September 2015, pp. 43–50 (2015)
2. Hussain, R., Nawaz, W., Lee, J.Y., Son, J., Seo, J.T.: A hybrid trust management framework for vehicular social networks. In: Nguyen, H.T.T., Snasel, V. (eds.) CSoNet 2016. LNCS, vol. 9795, pp. 214–225. Springer, Cham (2016). https://doi.org/10.1007/978-3-319-42345-6_19
3. Kooti, F., Lerman, K., Aiello, L.M., Grbovic, M., Djuric, N., Radosavljevic, V.: Portrait of an online shopper: understanding and predicting consumer behavior. CoRR, abs/1512.04912 (2015)
4. Lebedev, A., Lee, J.Y., Rivera, V., Mazzara, M.: Link prediction using Top-k shortest distances. In: Calì, A., Wood, P., Martin, N., Poulovassilis, A. (eds.) BICOD 2017. LNCS, vol. 10365, pp. 101–105. Springer, Cham (2017). https://doi.org/10.1007/978-3-319-60795-5_10
5. Lee, J.Y., Oh, J.C.: Agent perspective social networks: distributed second degree estimation. Encyclopedia of Social Network Analysis and Mining, pp. 1–12. Springer, New York (2016). https://doi.org/10.1007/978-1-4614-7163-9_110187-1
6. Lee, J.Y.: Reputation computation in social networks and its applications (2014)
7. Lee, J.Y., Duan, Y., Oh, J.C., Du, W., Blair, H., Wang, L., Jin, X.: Automatic reputation computation through document analysis: a social network approach. In: 2011 International Conference on Advances in Social Networks Analysis and Mining (ASONAM), pp. 559–560. IEEE (2011)
8. Lee, J.Y., Duan, Y., Oh, J.C., Du, W., Blair, H., Wang, L., Jin, X.: Social network based reputation computation and document classification. J. UCS **18**(4), 532–553 (2012)

9. Lee, J.Y., Lopatin, K., Hussain, R., Nawaz, W.: Evolution of friendship: a case study of MobiClique. In: Proceedings of the Computing Frontiers Conference, pp. 267–270. ACM (2017)

10. Lee, J.Y., Oh, J.C.: Estimating the degrees of neighboring nodes in online social networks. In: Dam, H.K., Pitt, J., Xu, Y., Governatori, G., Ito, T. (eds.) PRIMA 2014. LNCS, vol. 8861, pp. 42–56. Springer, Cham (2014). https://doi.org/10.1007/978-3-319-13191-7_4

11. (Lionel) Li, Z., Fang, X., Sheng, O.R.L.: A survey of link recommendation for social networks: methods, theoretical foundations, and future research directions. CoRR, abs/1511.01868 (2015)

12. Solonets, S., Drobny, V., Rivera, V., Lee, J.Y.: Introducing ADregree: anonymisation of Social Networks through Constraint Programming. Springer, LNCS (2017)

13. Szczepanski, P.L., Barcz, A.S., Michalak, T.P., Rahwan, T.: The game-theoretic interaction index on social networks with applications to link prediction and community detection. In: Proceedings of the Twenty-Fourth IJCAI 2015, Buenos Aires, Argentina, 25–31 July 2015, pp. 638–644 (2015)

14. Tigunova, A., Lee, J.Y., Nobari, S.: Location prediction via social contents and behaviors: location-aware behavioral LDA. In: 2015 IEEE International Conference on Data Mining Workshop (ICDMW), pp. 1131–1135. IEEE (2015)

15. Wu, L., Ge, Y., Liu, Q., Chen, E., Long, B., Huang, Z.: Modeling users' preferences and social links in social networking services: a joint-evolving perspective. In: Proceedings of the Thirtieth AAAI, 12–17 Feb, Phoenix, Arizona, USA, pp. 279–286 (2016)

16. Zhang, J., Tang, J., Li, J., Liu, Y., Xing, C.: Who influenced you? Predicting retweet via social influence locality. TKDD **9**(3), 25 (2015)

Author Index

Printed in the United States
By Bookmasters